HAYDN'S SUNRISE, BEETHOVEN'S SHADOW

HAYDN'S SUNRISE, BEETHOVEN'S SHADOW

Audiovisual Culture and the Emergence of Musical Romanticism

DEIRDRE LOUGHRIDGE

THE UNIVERSITY OF CHICAGO PRESS
Chicago and London

Deirdre Loughridge is a lecturer in the Department of Music at the University of California, Berkeley.

The University of Chicago Press, Chicago 60637
The University of Chicago Press, Ltd., London

Printed in the United States of America

25 24 23 22 21 20 19 18 17 16   1 2 3 4 5

ISBN-13: 978-0-226-33709-8 (cloth)
ISBN-13: 978-0-226-33712-8 (e-book)
DOI: 10.7208/chicago/9780226337128.001.0001

This book has been supported by the AMS 75 PAYS Endowment of the American Musicological Society, funded in part by the National Endowment for the Humanities and the Andrew W. Mellon Foundation.

Library of Congress Cataloging-in-Publication Data
Names: Loughridge, Deirdre, author.
Title: Haydn's sunrise, Beethoven's shadow : audiovisual culture and the emergence of musical Romanticism / Deirdre Loughridge.
Description: Chicago ; London : The Unviersity of Chicago Press, 2016. | Includes bibliographical references and index.
Identifiers: LCCN 2015046824| ISBN 9780226337098 (cloth : alk. paper) | ISBN 9780226337128 (e-book)
Subjects: LCSH: Music—18th century—History and criticism. | Music—19th century—History and criticism. | Mixed media (Music)—18th century—History and criticism. | Mixed media (Music)—19th century—History and criticism. | Music and technology—History—18th century. | Music and technology—History—19th century.
Classification: LC ML195 .L68 2016 | DDC 780.9/033—dc23
LC record available at http://lccn.loc.gov/2015046824

♾ This paper meets the requirements of ANSI/NISO Z39.48-1992 (Permanence of Paper).

CONTENTS

INTRODUCTION

Audiovisual Histories

HAYDN'S SYMPHONY NO. 6 BEGINS SOFTLY. THE FIRST VIOLINS play alone, *pianissimo*. Soon, they climb a few steps and are joined by the second violins a third below, also *pianissimo*. In the third measure, the remaining instruments enter in turn: first the low-register basses, celli, horns, and bassoon, then the middle-register violas and oboes, and finally a high flute. As the number of instruments increases, the orchestra begins a crescendo, each member playing gradually louder while the registral expansion continues. By the fifth measure, the sound has grown from the first violins' lone, *pianissimo* D to a tutti A-major chord spread over four octaves, *fortissimo* (example 0.1).

With the aid of the symphony's title—*Le matin* [The morning]—the orchestral crescendo becomes a vivid representation of sunrise. The representation works not only by tracing the ascent of the sun with its rising melodic trajectory, but also by means of the increasing registral spread and loudness: growing light becomes growing sound. For Haydn's employer, Prince Paul Anton Esterházy, the musical representation of sunrise likely activated a network of associations, the sun being a symbol of royal power and enlightenment.[1] More concretely, the slow orchestral crescendo might have called to mind ceiling paintings of a type found in the Great Hall at Esterházy's Eisenstadt palace, wherein the sun's splendor emanates from Apollo driving his sun chariot across the sky (figure 0.1).[2]

Compare this orchestral crescendo to another, composed half a century later: the transition from the third movement to the finale of Beethoven's Fifth Symphony. The accumulation is more protracted, less steady, the difference between its start and climax more extreme. After twelve measures of sustained ***ppp*** strings over the *pianissimo* thuds of a timpani, the second violins join the firsts and the dynamic notches up to *pianissimo*. The ensemble remains at this dynamic level for another twenty-nine measures as the first violins unfold a halting melody then gradually climb and slide

EXAMPLE 0.1. Haydn, Symphony no. 6 in D major/1, mm. 1–6

EXAMPLE 0.2. Beethoven, Symphony no. 5 in C minor/3, m. 324–iv, m. 4

from minor into major. As the violins reach their peak, the remaining instruments enter: first bassoons, then clarinets, and finally flutes, horns, and trumpets join timpani and strings to fill out a chord spread over nearly five octaves. Still, the dynamic marking increases only to *piano* until the last four measures, when with a final rush the full orchestra crescendos to the *fortissimo*, C-major theme of the finale (example 0.2).

Beethoven's crescendo is not only of a different magnitude than Haydn's, but also of a different order. The Fifth Symphony, of course, has no programmatic title, and for generations of listeners it has represented the pin-

EXAMPLE 0.2. *(continued)*

nacle of purely musical logic, combined with a sense of spiritual truth. Historians typically trace this way of apprehending the symphony to a review by E. T. A. Hoffmann, first published in 1810. For this critic, composer, and author of fantastic literature, Beethoven's Fifth Symphony exemplified the romanticization of music, which had finally left behind an attachment to definite objects, so as to open "to man an unknown realm, a world quite separate from the outer sensual world surrounding him"—a world of the in-

EXAMPLE 0.2. (*continued*)

finite.[3] Such an idealist conception of music had emerged in early Romantic writings of the 1790s, but Hoffmann supported it with an analysis of Beethoven's music, thereby articulating the arrival of a serious musical culture centered on composers' works faithfully performed for reverent contemplation in the concert hall.[4]

Yet this version of music history presents a narrow view of both Hoffmann's essay and the cultural developments in which it took part. For along

EXAMPLE 0.2. *(continued)*

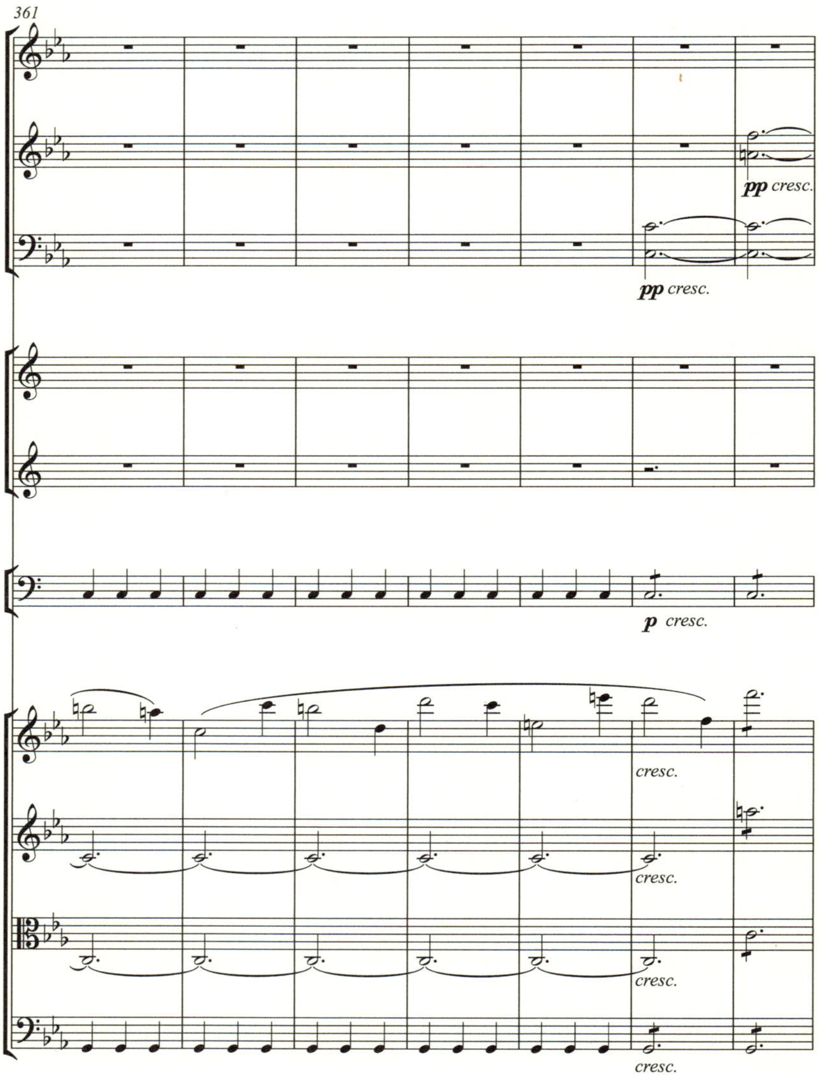

with invocations of the infinite and discussion of the musical notes, Hoffmann's review of Beethoven's Fifth provides a perceptual guide comparable to the title *Le matin*. The difference is that its referents come not from nature or its mythological interpretations, but from another world made sensory through the artful use of technology. The review is suffused with ghostly imagery—specifically the idea of "shadows that . . . draw closer and closer in upon us"—that conjures the phantasmagoria, a contemporary

EXAMPLE 0.2. (*continued*)

form of public entertainment in which a hidden image-projection apparatus made approaching ghosts seem terrifyingly real (a development to be explored in chapter 5).[5] The review thus suggests "seeing" Beethoven's crescendo as a ghostly shadow emerging from a great distance and finally looming toward one, the hushed opening indicating remoteness, the growing volume and registral spread increasing proximity. The effect relies not on a representation of light in sound but on an analogous technique for

EXAMPLE 0.2. *(continued)*

creating the illusion of three-dimensional motion in space, and with it a heightened sense of immersion in another world (figure 0.2).

This book is about the transition from Haydn's sunrise to Beethoven's shadow. During this period—roughly 1760 to 1810—optical devices such as magnifying instruments, peepshows, shadow-plays, and magic lanterns became widely available. They were the subjects of operas and popular scientific literatures, purveyed by street entertainers and scientific demonstrators, used at home and in public performances. These optical technologies did not only produce new forms of spectacle and practices of looking; they also gave rise to new spaces for music-making, practices of listening, and ultimately a new set of terms for describing and thinking about music. Through their particular processes of dissemination, these devices fostered the changes in musical perception illustrated by the comparison of Haydn's rising sun and Beethoven's approaching shadow. Whereas naked-eye observation of nature and painting provided primary reference frames for "seeing" the crescendo at the start of Haydn's Symphony no. 6 as a sunrise, the world-making powers of optical technologies framed otherworldly experiences of Beethoven's symphony. Whereas mimesis of nature and emotional expression furnished the main conceptual bases for making sense of music in the eighteenth century, notions of extending the senses and mastering invisible forces increasingly came to supplement or supplant them.

"[T]here can be no greater contradiction or clash in human cultures than that between those representing the eye and the ear."[6] So wrote Marshall McLuhan, whose historiographical model was one of great epochal shifts between the dominance of ear and eye: the printing press catalyzed a shift from ear- to eye-centric culture during the Renaissance, and the advent of electronic media in the twentieth century initiated a shift back to the ear. More recent research, especially in the field of sound studies, has countered the equation of modernity with visuality, demonstrating that hearing, no less than vision, was linked to reason and disciplined by technology.[7] Rather than assign modernity to one sense or the other, scholars have built from the thesis of a "separation of the senses," usually said to begin with physiological research in the early nineteenth century and to culminate in the technical media of gramophone and film.[8]

Still neglected, however, is the interdependency between sight and sound, looking and listening, and the ways interaction between the two perceptual modes has changed over time. Of course, work on various forms of multimedia—especially opera, film, and most recently digital media—

FIGURE 0.1. Paul Troger, "Apotheosis of Emperor Karl VI" (depicting Karl VI as Apollo driving the chariot of the sun). Ceiling fresco above the Emperor's Staircase, Monastery Goettweig, Austria, 1739. Photo credit: Erich Lessing/Art Resource, NY.

FIGURE 0.2. "Phantasmagorie Robertson," from Fulgence Marion, *L'optique* (Paris: L. Hachette et cie, 1867). Widener Library, Harvard University, KPD 4211.

has explored how visual and auditory elements come together to produce meaning in particular works and genres.[9] However, there has as yet been no effort to understand "audiovisual culture" comparable to those mounted on behalf of visual and auditory culture. As W. J. T. Mitchell has written, to call one's field of inquiry "visual *culture*" (as opposed to, say, "visual studies") is to posit that vision "is learned and cultivated, not simply given by nature; that therefore it might have a history related in some yet to be determined way to the history of arts, technologies, media, and social practices of display and spectatorship; and (finally) that it is deeply involved with human societies, with the ethics and politics, aesthetics and epistemology of seeing and being seen."[10] Analyses of multimedia "texts" often illuminate the interplay of multiple signifying systems, and the theories or ideologies that underlie their arrangement; but they rarely stray into the history of the senses, the conditions that enable certain audiovisual experiences, or the broader configuration of practices of looking and listening within a society.

An exception here is research on *grand opéra*, which has not only embraced the essentially audiovisual nature of the genre, but has also found fertile ground for investigating its technological and epistemological conditions in mid-nineteenth-century Paris. In the city Walter Benjamin dubbed the "capital of the nineteenth century," Anselm Gerhard identified urban realities that fundamentally altered the "perceptual structures" of its inhabitants. Daily experience of jarring juxtapositions, regular exposure to overwhelming representational techniques, and the instantaneities of electric telegraphy produced a new attitude toward contrast and an appetite for escalating spectacle that translated into operatic effects like the simultaneous electric-light sunrise and musical hymn in the act 3 finale of Giacomo Meyerbeer's *Le Prophète* (1849).[11] Sarah Hibberd has surveyed early nineteenth-century theories of auditory and visual perception, and para-theatrical spectacles such as panoramas, dioramas, and nocturnoramas, to show how these shaped a horizon of expectations for the creators and audiences for *grand opéra*.[12] John Tresch has shown Parisian physicists and artists in the mid-nineteenth century to have a common interest in developing new technologies to control light and sound, as well as a shared preoccupation with the unstable borderline between illusion and reality. Their pursuits found expression in spectacles like the diorama, the orchestral music of Hector Berlioz, the scientific lectures of Étienne Arago, and most fully in *grand opéra*. In such forms of performance, Tresch identifies "romantic audiovisuality," openly reliant on novel mechanical means to induce collective hallucinations and immersive experiences through music and spectacle.[13]

Scholars of music in turn-of-the-nineteenth-century Germany have conceived and approached their subject very differently. It was here that early Romantics began to write of "pure music," to occlude the material conditions of its performance, and to insist on its spiritual, otherworldly nature. The sources for early Romantic musical thought have been traced to the religious movement of pietism, which privileged the ear as the pathway to the soul, and to the philosophical movement of idealism, which developed an understanding of the arts as granting access to a higher realm of truth.[14] Emily Dolan has restored a crucial level of material culture to these efforts by relating eighteenth-century innovations in the design and use of musical instruments to a reevaluation of tone, which empowered early Romantics to prize "sound as sound," to regard the medium of music as inherently beautiful and capable of profound effects on mind and soul.[15]

The musical listening culture of late eighteenth- and early nineteenth-century Germany has taken on an almost mythical character for its supposed affinity for self-sufficient music and capacity for attention to the musical notes that went into decline with the perceptual changes epitomized by *grand opéra*.[16] Yet late eighteenth-century Germany had a thriving audiovisual culture, engaged with technological manipulations of sight and sound and promoting processes of cultivation between and across vision and hearing. Foundational texts of musical romanticism evidence this audiovisual culture; indeed, it was this culture that nurtured new modes of musical listening, and from which music emerged as the most metaphysical and otherworldly of the arts.

We can begin to unpack the relationship between audiovisual culture and musical romanticism by rethinking the "separation of the senses," a discursive project underway already in aesthetics prior to its heralded arrival in physiology. In the mid-eighteenth century, aesthetic theorists—newly conceiving their discipline as a study of "how things are to be known by means of the senses"—turned attention to the nature of vision and hearing.[17] Here, they found absolute difference rooted in the categories of space and time: visual arts were arranged in space, and musical and poetic arts took place in time. As Johann Gottfried Herder proclaimed, "space cannot be turned into time, time into space, the visible into the audible, nor this into the visible; let none take on a foreign field, but let it rule in its own the more powerful, the more certain, the more noble."[18] Such thinking laid the theoretical ground for medium purity, arts that ministered to the strengths and limitations of a single sense being privileged over hybrid forms such as opera and dance.[19]

Rather than rendering the visual and auditory independent fields of operation, however, the "separation of the senses" raised new questions about audiovisual relations, and multiplied the possibilities for manipulating and experiencing their interaction.[20] Following his remarks about the absolute difference between the visible and the audible, Herder observed that their fields nonetheless inevitably come together in the human perceiver:

> Precisely because the arts in respect of their medium exclude one another, they secure their realm; united nowhere but in the nature of man, in the midpoint of our sensation. How this is to enjoy and arrange them depends on our taste, or much more on our ordering reason.[21]

Herder, in other words, proceeded from the "nature" of the audible and visible to the "culture" of the audiovisual: sight and sound come together in human experience, in ways subject to the historically variable states of taste and reason.

Much attention has been paid to the ways in which early German Romantics maligned the visual. Writers like Herder, Jean Paul, and Wilhelm Wackenroder advocated a new type of listening that—in contrast to the primarily sociable modes then prevalent—required shutting one's eyes to the surrounding world and giving total focus to the music. They criticized "tone-painting," the attempt to represent visible objects in tones; and they disparaged the ocular harpsichord, an instrument that was supposed to convert tones into colors so as to create visual music. These threads of early Romantic discourse have supported musicological accounts of Romanticism as privileging sound (or sounding structure) alone, to be worshipped as the idealized object of attentive listening.

Yet even as they policed a boundary between the musical and the visual, early Romantics perceived music audiovisually; in fact, the conceit of "pure music" made the power of tones to stimulate images a source of wonder. Wackenroder captured the sense of contradiction between the idea of "pure music" and the audiovisual impression it made in practice with the significant qualifier "and yet" in the following passage:

> Ideal, angelically pure art knows in its innocence neither the origin nor the goal of its excitations, does not know the relationship of its emotions to the real world. And yet, despite all its innocence, through the overwhelming magic of its sensual force it arouses all the wonderful, teeming hosts of the fantasy, which populate the musical strains with

> magical images and transform the formless excitations into distinct shapes of human emotions, which draw past our senses like elusive pictures in a magical deception.[22]

While the first part of this passage encapsulates the idealist aesthetics typically associated with early Romantics—music belonging to a higher reality independent of earthly phenomena—the swerve "and yet" points toward a different understanding: its causal links between music's sensual force, the listener's fantasy, and magical images describe an audiovisual conception of music grounded in the listener's body and comparable to contemporary image-projection apparatuses ("like elusive pictures in a magical deception").[23] Ludwig Tieck detected a similar rift between the idea of pure music and the experience of listening. Comparing symphonies to dramas, he wrote that orchestral works

> envelop the greatest enigma in enigmatic language, they depend on no laws of probability, they need connect to no history and no character, they remain in their purely poetic world. Thus they eschew all means to draw us in, to enrapture us; they are their subject-matter, from beginning to end: the goal itself is present in every moment, and begins and ends the work of art. And yet often such individually-graphic images swim in the tones, so that this art, I'd like to say, captures the eye and ear at the same time.[24]

Like Wackenroder, Tieck marked the mysterious ability of music, an art seemingly divorced from the world and addressed only to the ear, to stimulate the image-forming capacities of the mind and thereby capture the eye as well. The ocular dimension of these writings points to transformations in the "perceptual world" of the eighteenth century, having to do not only with musical instruments in the ways Dolan has demonstrated but also with optical technologies and their uses.[25]

In his history of the senses, Robert Jütte notes the difficulty of distinguishing between the historicity of perceptions and the forms in which they have been recorded, especially when it comes to telling mere changes in discourse and its rhetorical figures apart from transformations in the ways the senses are used.[26] But while Jütte presents perceptual experience and discourse as separate registers, other scholars have shown how the two are in fact interdependent. Katherine Hayles, building on Mark Johnson's work on bodily experience as the underpinning of language, theorizes the relationship between discourse and bodily practices in terms of feedback loops: "when people begin using their bodies in significantly different ways, either because of technological innovations or other cultural shifts," she writes, "changing experiences of embodiment bubble up into language, affecting the metaphoric networks at play within the culture. At the same time, discursive constructions affect how bodies move through space and time, influence what technologies are developed, and help to structure the interfaces between bodies and technologies."[27] Technology plays a central role in this account, not as a unique or autonomous driver of change, but because it represents one source of new bodily experiences that find expression in language and alter the connections among cultural phenomena—a source especially salient for both the emergence of the posthuman of concern to Hayles, and the emergence of musical romanticism of concern in the present book.

One way I approach texts in this study, then, is to mine them for shifting patterns of metaphor, so as to find in language the traces of changing experiences of embodiment. Texts also play another role, as mediators of the audiovisual. While audiovisual recording media allow analysis of fixed music-image relations, the sounds of audiovisual forms such as peepshows, magic lantern shows, and shadow-plays require imaginative reconstruction from written sources, and determining the coordination among speech, music, and image—even when abetted by surviving artifacts—likewise relies on written testimonies. Besides supplying evidence for oral performances, texts can also induce audiovisual perceptions. Current scholarship tends to use "audiovisual" to contrast sound-and-image media from textual media, but in the eighteenth and nineteenth centuries this line was not so bright. Theorists of the time were alert to the varied capacities of media to produce sensory experiences beyond their own material qualities—to stimulate the imagination. According to Jean-Jacques Rousseau, music and poetry shared the ability to portray "everything that the imagination can envisage," unlike painting, which could address only the eye.[28] To study audiovisual culture in this period is thus not to exclude language and verbal discourse, but to take seriously the ways in which they—like music—could produce audiovisual experiences as vivid as those of a magic lantern show; and additionally, to take seriously the implications of such resemblance when it was noted.

From our current perspective, this book brings together "audiovisual culture"—found in technology-driven forms of opera, popular science, and street entertainments—with "musical culture"—represented by works like

Haydn's *Creation* and Beethoven's instrumental music—in order to illuminate Romantic listening practices and aesthetics. But it also aims to historicize perceptions of a division between audiovisual and musical culture. Socially and institutionally, musical and audiovisual forms mixed at the turn of the nineteenth century. The Vienna premiere of Beethoven's Piano Concerto no. 5, for example, shared its program with tableaux vivants, people posing in costume with scenery and lighting to recreate famous paintings, between its movements.[29] Far from possessing a vaunted history as the art toward which all others aspire, moreover, music—especially without words—was of dubious standing. From the warnings Georg Philipp Telemann received against a musical career ("I would become a traveling entertainer, tightrope walker, minstrel, squirrel master, etc. if music were not taken away from me"), to G. W. F. Hegel's contention that music was not properly considered an art ("because the one chief thing in all art, namely spiritual content and expression, is missing from it"), the precarious position of music is evident: it was easily lumped with merely sensory pleasures, excluded from the sphere of cultural products capable of moral uplift and worthy of intellectual (as well as capital) investment.[30] As David Gramit and Celia Applegate have shown, this was a time when musicians sought to secure esteem for their profession, in part by arguing for the value of music to the general cultivation of the individual.[31] It was also by selectively likening music to or distancing it from certain optical technologies and audiovisual forms that proponents of a "culture of serious music" staked out their terrain. In other words, the ability to separate music from its visual and technological conditions (whether mentally, socially, or architecturally) is a historically contingent facet of audiovisual culture—and early nineteenth-century audiovisual culture is marked by budding efforts to develop that ability in face of the prevailing mixtures of performance and experience.

Approaching musical works as part of audiovisual culture rather than within audiovisual contexts further makes room for their active role in the history of looking and listening. As Rita Felski has argued, situating a text in a context tends to cast the former "only as an object-to-be-explained rather than a fellow actor and cocreator of relations, attitudes, and attachments" (an effect evident in Gerhard's study of *grand opéra* in urban context).[32] The crescendos discussed above are not so much explained by contemporaneous visual forms as they are part of a field of practices that involved ear, eye, and the interactions between them, and that cultivated certain dispositions in their perceivers. Consider in this regard another crescendo, from Mozart's opera *Idomeneo* (1781). Where the libretto indi-

EXAMPLE 0.3A. Mozart, *Idomeneo* no. 14, Marcia, mm. 1–6

cates that "a harmonious march is heard in the distance," Mozart provides a march scored for muted instruments playing *piano* or *pianissimo* (example 0.3a). Over the muted march, Electra sings, "I hear from afar a melodious sound which calls me to go aboard; it is time to go." She exits in haste, and "the march is heard ever closer as the scene is changed" to the port of Kydonia—an effect Mozart achieves by means of increasingly full orchestration, a staggered removal of mutes, and a crescendo to fortissimo (example 0.3b).

Music histories have little to say about the "march heard ever closer" crescendo. It is only with *opéra comique* after the French Revolution that sophisticated instrumental effects are considered to come into their own, the elaborate deployment of spatialization and tone-modifying devices preparing the way for Romantic opera and orchestral music.[33] Nor does the crescendo contribute to the history of listening in the ways James Johnson has identified for the operas of Gluck and Rossini. Whereas Gluck prompted "audiences to transcend the imitative model of listening" by eschewing depictions of natural objects in favor of clear harmonic-rhythmic

EXAMPLE 0.3B. Mozart, *Idomeneo* no. 14, Marcia, mm. 25–32

All' ultima ripetizione di sopra, facendo pausa, i corni e le trombe ritirano li sordini, ma suonano molto piano. L'orchestra va crescendo sino al fortissimo.

On the final repetition from above, taking a pause, the horns and trumpets remove their mutes but play very quietly. The orchestra makes a crescendo to fortissimo.

delineations of emotional nuances, Rossini, through vocal pyrotechnics and orchestral effects disconnected from representation, "made audiences listen to music, not as imitation or image or emotion, but as sheer music."[34]

From the perspective of audiovisual history, however, Mozart's crescendo acquires significance. Though its technique for making the march sound "ever closer" may now seem tame and obvious, at the time it was not. Spatial effects in opera were nearly always obtained by repositioning musicians in the wings or backstage (muted instruments, as we shall see in chapter 1, were reserved for expressing soft or sad phenomena, and typically involved only strings). To instead mute the orchestra required special equipment: Wolfgang wrote from Munich to his father in Salzburg to send a "*trumpet mute*—of the kind we had made in Vienna—and also one *for the horn*." Unable to obtain the requested horn mute, Leopold reassured his son that "the main thing is anyway only the trumpet mutes, which is something strange and new."[35]

The difference between relocating and muting the instruments was not only logistical but also conceptual, requiring a shift from imitating the sound source—the cause of the distant effect—to imitating the effect alone. The crescendo thus asks opera spectators to accept the illusion of distant instruments and motion toward them in lieu of a physical reconfiguration of their environment. In other words, the number calls for a form of spectatorial consciousness that suppresses awareness of the total theatrical space in exchange for the illusion of motion through the fictional world; a novel instrumental effect serves to heighten the experience of transport. Framed this way, the "ever closer" crescendo participates in the cultivation of what Anne Friedberg has called the "mobilized virtual gaze," a perceptual mode she identifies with visual apparatuses (such as the panorama, diorama, and ultimately cinema) designed to transport their spectators.[36] The crescendo also shares its technical innovation with the "phantasmagoria": like the phantasmagoria, it achieves its impact by obscuring the means of production and by creating an illusion of motion.

Mozart's "ever closer" crescendo likewise figures in the transition from Haydn's sunrise to Beethoven's shadow. It suggests—as this book will trace in more detail—how visual technologies and musical works collaborated in training listeners to become not only more attentive but also more fully immersed in other worlds and the apparent reality of those worlds. To date, histories of listening have been concerned with shifts between a certain set of parameters: distracted/attentive, noisy/silent, passive/active, detail/whole, associative/structural. But listening—no less than spectatorship—can be analyzed in terms like those proposed by Thomas Elsaesser for multimedia, terms that include the fixity or mobility of the perceiver, his or her distance from or proximity to the perceived, and the impression of looking through a frame versus being completely immersed in another space.[37] These latter parameters have been considered mainly in relation to visual apparatuses, but they are equally subject to auditory manipulation. While terms like *listener* and *spectator*, *musical work* and *visual technology*, cannot be abandoned without serious linguistic contortions, they should be understood to reflect not states of sensory separation or medium purity, but rather different emphases within audiovisual situations.

As the boundary points "Haydn's sunrise, Beethoven's shadow" suggest, this study is centered not only temporally in the late eighteenth and early nineteenth centuries, but also geographically on German-speaking lands. It is the perceptual world of early Romantics—of figures like Wackenroder, Tieck, and Hoffmann, credited with inaugurating attentive listening and idealist music aesthetics—that I am primarily concerned with recovering.

Like the philosophical writings of Kant, early Romantic writings on music are suffused with references to optical technologies that have gone unrecognized because the structures of experience they invoked, the metaphorical networks in which they participated, became obsolete for readers.[38] By focusing on Germany, we thus stand to gain understanding not only of its audiovisual culture, but also of how its audiovisual culture shaped the listening practices and aesthetics of musical romanticism.

Although Germany provides a focus for this study, developments there did not take place in isolation. Indeed, the forces that spurred the rise of optical devices—the proliferation of popular science texts, the promotion of edifying entertainments, and related societal changes conveniently covered by the term *Enlightenment*—applied across Europe. Often discourses and practices around optical instruments originated elsewhere and reached Germany through texts or traveling performers. Operas from Italy and France brought opportunities for staging televisual technologies, for instance, while literature aimed at popularizing natural philosophy and its instruments came from France, Italy, and England before finding local counterparts (see chapter 1). Savoyards—from the French-Italian region of Savoy—traveled throughout Europe, establishing pan-European forms of audiovisual entertainment even as they remained outsiders, their contributions viewed with suspicion (see chapters 2 and 4). Further, the relationship of German to non-German cultures was a prominent concern for intellectuals in this period of cultural identity formation. Thus, French sources could provide direct models for emulation (for shadow-plays, for example, discussed in chapter 3) or for repudiation (as in the case of the nocturnorama, discussed in chapter 4). The flows of people, printed matter, and artifacts across Europe were crucial to conditions in Germany, and though these cannot always be retraced with precision, I have made an effort to register them where relevant to the development of German audiovisual culture.

This book does not offer an exhaustive account of audiovisual forms that flourished in the period from Haydn to Beethoven. It mostly leaves aside tableaux vivants and panoramas, for example, although they too emerged at this time, and much has been written about their implications for perception and connections to nineteenth-century operatic spectacle.[39] The selection of materials, like the geographical focus, has been guided by the concern with the emergence of musical romanticism: each chapter centers on a visual technology that came to play a significant role in musical discourses that promoted attentive listening to music as essentially otherworldly. These technologies prove to be of two types: those involving peep-lenses (telescopes, microscopes, peepshows) and those involving projected images (shadow-plays, magic lanterns). All were subjects of the field of optics, and by the 1770s were both discussed in popular scientific literature and available for purchase and use in the home.

This is not to say that spectacles like the panorama were less technological than those involving lenses, but that their technologies and cultural dissemination differed in ways that made them less formative for early Romantic musical thought. Whereas the panorama aimed at precise reproduction of the natural world, peep-lenses and projected images involved transformations of perception that tended to the otherworldly via magnification or phantasmatic appearances. This made them more conducive or comparable to otherworldly modes of listening—modes that also drew support from eighteenth-century religious and philosophical currents such as pietism and idealism. Thus, to recall Hayles's model of mutual influence between technology and discourse, the book's coverage reflects the fact that new visual technologies did not simply make people listen to or interpret music differently; rather, discourses also shaped how people experienced those technologies, and the connections they drew between them and other aspects of their world.

The chapters that follow trace developments from the eighteenth century through the first decade of the nineteenth century multiple times: each examines an optical technology and its associated sights, sounds, and meanings across this period, considering these in relation to bodily and mental experience and following their movement from material practice into musical discourse. The first chapter, "From Mimesis to Prosthesis," centers on magnifying instruments and the cross-sensory analogies that cultivated a "prosthetic mode" in music and listening. Operas and philosophical texts elaborated two different musical analogues to magnifying instruments: muted tone and keyboard improvisation. These analogies also likened peeping and eavesdropping, and made desirable the position of secret, outside observer. Crucially, the valence of these analogies depended on a change in attitude toward magnifying instruments: in the latter half of the eighteenth century, popular scientific literature promoted telescopes and microscopes as tools of sensory extension and knowledge rather than of sensory deception and blindness. With this change, magnifying instruments became a condition of possibility for characteristically Romantic approaches to listening and musical expressions of the otherworldly—approaches the chapter illustrates with the slow movement of Beethoven's Piano Concerto no. 5.

The peepshow shared with magnifying instruments the interface of the peep-lens. But as a series of pictures presented by itinerant Savoyards, the peepshow offered "new worlds" to view as cheap entertainment rather than knowledge. Whereas the peepshow has been studied as a visual apparatus, my second chapter recovers its auditory dimensions, and its audiovisual dynamic of "listening toward looking." Both the sounds and the hidden sights of the peepshow were important considerations for its operatic treatments. From the 1770s to 1790s, peepshows were a popular topic of German operas. The hidden spectacle of the peepshow allowed composers to explore how to make music and words point to or produce visual experiences, at a time when keyboard-vocal scores meant opera was increasingly consumed away from the operatic stage. As the broad accessibility sought for German opera in the late eighteenth century gave way to early Romantic ideals of high art, however, the proximity between peepshow and opera became undesirable, and the device came to figure the trivial "series" of elements that German opera had to leave behind in order to transport spectators into a unified other world. Yet, even as the lowly status of the peepshow contributed to hardening distinctions between high art and mere entertainment, the audiovisuality of the peepshow—the disposition of listening toward looking—survived as a Romantic mode of musical engagement.

Chapters 3 and 4 turn to the image-projection apparatuses of shadow-play and magic lantern. Chapter 3 uses a 1781 shadow-play performance of Johann André's through-composed ballad "Lenore" to explore late eighteenth-century interactions between poetry, music, and moving images, as well as their contribution to the rising status of through-composed song. The shadow-play helped teach a new mode of listening to lieder by directing listener focus away from the elements of composition and performance—where it normally centered—to the audiovisual experience these might conjure for the imagination. But such presentations raised questions about what performance conditions and musical techniques were best calibrated to foster imaginative listener engagement. These questions became particularly acute in the early nineteenth century with the reception of Haydn's *Creation*, an oratorio that critics likened both to a shadow-play and to a magic lantern. Chapter 4 discusses how these two entertainments furnished different models for perceiving the interaction between words, music, and visuals in *The Creation*, and with them different aesthetic evaluations of the work, and different stances on the compatibility—or lack thereof—between great works of music and moving images.

The next chapter returns to E. T. A. Hoffmann's review of Beethoven's Fifth Symphony. As mentioned above, musicologists have come to see E. T. A. Hoffmann's review of Beethoven's Fifth Symphony as establishing the foundation for serious musical cultures in the nineteenth and twentieth centuries. In casting Hoffmann as a modern musical thinker, however, scholars have overlooked alternative aspects of his writings, such as his remark that in listening to Beethoven's instrumental music we become "spirit-seers" (*Geisterseher*).[40] The term "spirit-seer" points to a set of philosophical debates, literary works, and performance practices centered on the question of whether it is possible to perceive ghosts—a question that transferred uncertainties about sensory deception and extension from optical technologies to the imagination and "inner" senses. Chapter 5 traces the development of audiovisual strategies for inducing ghost belief and rational strategies for defending against such belief, the interaction of which culminated in the form of ghost show called phantasmagoria. Through the image of approaching shadows, Hoffmann's review links Beethoven's Fifth Symphony to the key technological innovation of the phantasmagoria; this connection in turn recasts his discussion of underlying rational coherence—celebrated for heralding a new focus on musical structure—as a means to prove that the spirit-seeing induced by the symphony consisted not in wild imaginings or optical deceptions but rather in metaphysical truth.

Because this book deals with the music of peep-lens and image-projection apparatuses, there will be much in its pages for those interested in the "prehistory" of cinema or film music. Indeed, I have benefited greatly from research on so-called "pre-cinema," which has done much to recover and explicate apparatuses that predate Edison's and the Lumière brothers' inventions. But I resist framing this study as a "prehistory," or describing its objects as "proto-cinematic." Other musicologists have taken this approach. Anno Mungen's "archaeology of film music," for instance, examines the music for panoramas, dioramas, tableaux vivants, and magic lantern shows so as to demonstrate its similarities to and influence upon cinema. In examining the writings of early Romantics such as Wackenroder, meanwhile, Peter Franklin identifies "narrativizing responses to music" that were "'proto-cinematic,' not least for their richly visual character."[41] Rather than think about such audiovisual forms and discourses in terms of what they were not (yet), I am concerned with thinking about them in terms of what they were—to recover the meanings and interconnections that pertained for those who used and experienced them in the eighteenth and early nineteenth centuries. Wackenroder's contemporaries could not have found his accounts of listening "proto-cinematic," but they could very well have found them magic lantern–like—an association of both more dubious and atavistic significance (as we shall see in chapters 4 and 5).

Studying the audiovisual culture of the eighteenth and nineteenth centuries yields new relations between past and present, a historiographical matter to be revisited in the concluding chapter ("Audiovisual Returns"). Walter Benjamin wrote that no fact is historical a priori; rather, it becomes "historical posthumously, as it were, through events that may be separated from it by thousands of years." The historian who adopts this perspective, rather than seeking to recount a continuous sequence of events, "grasps the constellation which his own era has formed with an earlier one."[42] Scholarship of the late twentieth century positioned Hoffmann within teleological narratives of progress toward pure or "absolute" music, reading his review of Beethoven's Fifth Symphony as a key step toward modern valuations of organic unity, and the conviction that "to understand music means . . . grasping the structure, the harmonic and thematic logic of a work."[43] The present study discovers our commonality with Hoffmann in another way: in powerful musical experiences shaped by encounters with visual technologies. However, I also aim to recover the otherness of the early Romantic era. Riven by a genuine uncertainty about the possibility of seeing ghosts, and hence about the spiritual status of inner visions, the perceptual world of early Romantics was profoundly "other" in ways that teleological narratives have failed to register. In the balance of self-recognition and difference, of shared and alternate realities, the period from Haydn to Beethoven finds a meaningful constellation with our own.

This book is ultimately about changes that took place around the turn of the nineteenth century, a time often described in terms of great social, political, economic, and intellectual upheavals that converged in the radical break with the past called "modernity." But this history of audiovisual culture is not an account of rupture or paradigm shift. Rather, it is an account of diffusion, accumulation, and transition—of change with continuity. Whereas narratives of rupture or paradigm shift tend to mute competing modes of perception or interpretation that coexist in time, the chapters that follow highlight such instances, finding in them clues to how the transition from Haydn's sunrise to Beethoven's shadow took place. This is not simply a story of new technologies introducing new perceptual experiences, bodily practices, and conceptual possibilities; it is also one of people negotiating the meanings and implications of their audiovisual world.

ONE

From Mimesis to Prosthesis

In 1777, Haydn composed an opera based on Carlo Goldoni's libretto *Il mondo della luna* (1750). The opera tells a familiar *buffa* tale of a father thwarting his daughters' romantic happiness, but the intrigues that lead finally to love-based marriages stem from inquiries into the world of the moon. The opera begins on the astronomer Ecclitico's roof, where four students prepare to carry his new, giant telescope up into his observatory tower to discover if the moon is filled with people. Just then, Buonafede—an amateur astronomer and father of two lovely daughters—arrives, troubled that he can find "no theory explaining what the moon is." Ecclitico tells Buonafede that the moon is a transparent body inside of which there is another world, and when Buonafede asks how he knows this, Ecclitico explains: "I have built a telescope which pierces through so much that it offers a view of both the surface and the core. Not only does it show kingdoms and provinces, but also houses, squares and people. With my big telescope I can see up there, to my pleasure, the women undress themselves before going to bed." Eager to see for himself, Buonafede enters Ecclitico's observatory, looks through his giant telescope, and spies lunar men and women engaged in various private activities. Little does he realize, the telescope is trained not on the distant moon but on an earthly machine. When Buonafede enters the observatory, Ecclitico commands his servants to "move the machine, drag it near the telescope so that, by looking through it, Signor Buonafede thinks he sees every figure move in the World of the Moon." Meanwhile the opera audience, according to Goldoni's stage direction, sees "an illuminated machine drawn to the top of the telescope, inside of which some figures are moving."[1]

Goldoni designed this observatory scene such that while Buonafede looks through the telescope, the false astronomer addresses the audience in recitative, mocking Buonafede's credulity (his name means "good faith")

EXAMPLE 1.1. Haydn, *Il mondo della luna*, act 1, scene 3, Intermezzo no. 1, mm. 1–10

with remarks such as "Buonafede now believes he sees lunar women up there, while lunar women are down here."[2] When Haydn set the scene to music, however, he altered Goldoni's design by adding instrumental numbers. As Buonafede views a young girl caressing an old man, the audience hears not Ecclitico's commentary but rather a brief, wordless intermezzo. With its stilted and repetitive melody and bass line tick-tocking between scale degrees one and seven, the opening phrase of the intermezzo suggests the mechanical movements of the figures in the false moon. Delicately scored for *divisi* violins with mutes, the number also sounds distant and extraordinary, inviting us to believe with Buonafede that we are truly peering into another world (example 1).

Haydn's transformation of Goldoni's observatory scene registers a growing acceptance of magnifying instruments outside the narrow circles of natural philosophers—a shift from suspicion to enthusiasm for their ability to reveal other worlds. It also opens onto connections that began to take shape in the 1770s between magnifying instruments, musical techniques, and their associated modes of looking and listening. As Jonathan Sterne has noted, media technologies are often assumed to have either a visual or auditory nature, but there is much to be learned by setting aside such classifications to examine the ways the senses have been used "through and around media."[3] While the magnifying instruments of the laboratory, drawing room, and parterre extended vision alone, the operatic stage and philosophical and fictional literatures forged analogies between magnifying instruments and artificial extensions to hearing. Haydn's observatory scene demonstrates one such analogy in its link between telescopic looking and muted tone. Another was obtained between magnifying instruments and keyboard improvisation, which were likened as means to reveal the otherwise imperceptible, and were said to position the observer as a secret peeper or eavesdropper. At a time when musical listening was generally regarded as an interactive, sociable activity, analogies with magnifying instruments helped familiarize a covert, outside position for the listener. Additionally, the prosthetic power of magnifying instruments provided an alternative to the aesthetic frameworks of mimesis and expression, equipping listeners to accept what they found unfamiliar and incomprehensible as contributing to their knowledge of worlds beyond the unaided senses.

The prosthetic mode—being both a listening attitude and musical techniques able to call forth that attitude—provided a historical and conceptual intermediary between mimesis and metaphysics: by mimicking this-worldly technologies, muted tone and keyboard improvisation suggested access to other worlds. These acoustic analogues to magnifying instruments—these musical forms of sensory extension—came together in the slow movement of Beethoven's Piano Concerto no. 5, where their prosthetic resonances have since dissipated to leave present-day critics with general impressions of intimacy and a higher reality. A precondition for the prosthetic mode, however, was not merely the invention of magnifying instruments but also their popularization as instruments of knowledge—a cultural process that Goldoni's *Il mondo della luna* and its setting by Haydn allow us to trace.

Magnifying Instruments

From their introduction in the seventeenth century, telescopes and microscopes were conceived as what we would now call prostheses: they extended vision to the distant and to the small, revealing not only new objects—or "new worlds," as was often said—but also a deficiency in the God-given eye.[4] Thus, Robert Hooke wrote of the senses suffering from "infirmities" that could be rectified by "the adding of *artificial Organs* to the *Natural*," and sought "to promote the use of Mechanical helps for the Senses, both in the surveying the already visible World, and for the discovery of many others hitherto unknown."[5] The use of instruments to reveal unknown phenomena to the senses was novel, and differentiated the emerging brand of natural philosophy—pursuing knowledge of nature through experiment and observation—from mathematical sciences, with their long-standing use of measuring tools for largely practical ends.[6] Today, the notion of prosthesis not only finds ready application to magnifying and other scientific instruments that expand the range of perceptible phenomena, but also reigns as a paradigm for thinking about media in general, which through various interfaces with the body function to extend the action of speech, memory, and other human capacities across space and time.[7]

But if the telescope was emblematic of sensory extension, it was equally emblematic of sensory deception. Before being applied to the study of nature, magnifying lenses were devices of "natural magic," their main purpose being to produce wondrous effects by hidden means—to trick the senses.[8] Well into the eighteenth century, magnifying instruments retained strong associations with illusion and error. The dubious status of the telescope finds expression in the frontispiece to Laurent Bordelon's *L'histoire des imaginations extravagantes de Monsieur Oufle* (1710), where it is allied with conjurors, tricksters, and the "extravagant imaginings" of the story's eponymous character (figure 1.1). For the French philosopher Claude Adrien Helvétius, the telescope furnished a ready illustration of human susceptibility to illusion. In his treatise *De l'esprit* (1758), Helvétius recounted the "well known story of a country clergyman and an amorous lady." Having heard that the moon is inhabited, the lady and clergyman try to view the lunar people through their telescopes. The lady sees two shadows inclining toward each other and identifies them as happy lovers. The clergyman too sees a pair of shadows, but dismisses her interpretation: clearly, the two shadows are the two steeples of a cathedral.[9] The telescope features here as an instrument through which people see what they wish to see, finding not another world but themselves reflected back.

The satirical popular culture of the seventeenth and eighteenth centuries fueled such views of magnifying instruments, providing a site of critical resistance to the claims of natural philosophy to reveal nature's secrets through instrumental practice.[10] It was in this popular culture that Goldoni's *Il mondo della luna* had its roots. At the suggestion of his patron Count Grosberg, who remembered the piece from the "old fair of Paris,"

FIGURE 1.1. Laurent Bordelon, *L'histoire des imaginations extravagantes de Monsieur Oufle*, frontispiece (Amsterdam: Estienne Roger, 1710). Rare Books Division, Department of Rare Book and Special Collections, Princeton University Library, 3236.63.346.

Goldoni adapted the story of a *comedia dell'arte* piece first performed at the Comédiens Italiens du Rois in Paris in 1684: Nolant de Fatouville's *Arlequin, Empereur dans la Lune*.[11] Like Goldoni's, Fatouville's comedy concerned a moon-world hoax. It too began at a telescope, as Pierrot and a Doctor debate the existence of a world on the moon. But the telescope in

Fatouville's scenario provided only a backdrop, an occasioning factor for the philosophical debate and subsequent hoax perpetrated against the Doctor. The telescope did not entail stage machinery; and it was not employed in the ruse, which instead involved a false lunar visitor to Earth who described his fantastic world to the Doctor.

Goldoni's expanded role for the telescope allowed not only for the spectacle of an illuminated machine, but also for greater dramatic development of the issues attending telescopic vision. Buonafede's experience illustrates a number of themes common to depictions of magnifying instruments, among them the telescopic viewer's blindness to the world around him; his curiosity or desire for knowledge; and his status as a secret observer. For comic writers like Goldoni, as well as for philosophers who distrusted optical instrumentation, these themes furnished material for satire and critique. But as Haydn's setting of *Il mondo della luna*, and subsequent invocations of magnifying instruments in musical contexts, suggest, they could also be reinterpreted as the necessary conditions for genuine contemplation of other worlds.

The pompous philosopher blind to the goings-on around him was a popular stereotype of eighteenth-century *opera buffa*, invoked in *Il mondo della luna* when Buonafede's daughter Clarice remarks about the astronomer Ecclitico, "as long as he thinks about the moon and the sun, his wife will be free to do whatever she wants."[12] As Catherine Wilson has noted, the "absorption of the learned in their subject is always potentially comic (because of all that they do not see)."[13] Magnifying instruments, however, represented a heightened form of absorption and blindness, for they literally altered what the eyes could take in, excluding the surroundings and tuning them to phenomena on a different scale. According to critics, this sensory tuning did not bring observers closer to knowledge but rather removed them from the world with which they should be concerned. As Patrick Singy has shown, many eighteenth-century philosophers were suspicious of differentials in sensory acuity, instead prizing the very commonality that allowed for agreement among different observers, and which thereby furnished a firm ground for knowledge.[14] This attitude also had an ethical dimension, for to go beyond the God-given sensory apparatus was also to upset the natural order. Alexander Pope expressed this ethic in his internationally read *Essay on Man* (1734):

> The bliss of man (could pride that blessing find)
> Is not to act or think beyond mankind;

No powers of body or of soul to share,
But what his nature and his state can bear.
Why has not man a microscopic eye?
For this plain reason, man is not a fly.[15]

Pope's poem gave sober voice to what theater and opera expressed through comedy: that those who seek knowledge of all God's creation are fools; the wise remain in their God-given sphere.

Goldoni's telescope scene offered a similar critique of science: if Buonafede had looked around with his own eyes rather than through the telescope, he would have seen the false lunar world and the trick. One of numerous anonymous rewrites of Goldoni's libretto, this one set by Gennaro Astarita for Venice in 1775, sharpened the critique. Goldoni's original choral finale related only obliquely to scientific inquiry, instead offering a chastising lesson about social order: "This is what happens to people who want to change their fortunes: they hope and believe in everything; in the stars and in the moon. But he who wants to become a lunatic will regret it in the end."[16] The final chorus of Astarita's version, by contrast, reiterated the folly of the natural philosopher: "The world of the moon looks beautiful to the ignorant, but people with brains stay in this world."[17]

Magnifying instruments were thus tied up with broader debates about the scope and means of human inquiry, at a time when the desire for knowledge—curiosity—was seen as highly problematic. In his intellectual history of the modern age, Hans Blumenberg credits Galileo's telescope with initiating the "unfettering of curiosity," its special influence among emerging scientific methods owing to its adaptability to "the classical ideal of the contemplation of nature," by contrast with experimental apparatuses that intervened in nature's operations.[18] Early proponents of magnifying instruments indeed identified the advantages of distanced observation over active intervention. As Robert Hooke explained in 1665, contrasting the messy vivisectionist with the stealthy microscopic viewer:

> when we endeavor to pry into [nature's] secrets by breaking open the doors upon her, and dissecting and mangling creatures whil'st there is life yet within them, we find her indeed at work, but put into such disorder by the violence offer'd, as it may be easily imagined, how differing a thing we should find, if we could, as we can with a *Microscope* in these smaller creatures, quietly peep in at the windows, without frightening her out of her usual bays.[19]

Yet while Hooke meant to reflect positively on the microscopic observer's covert, outside position, his account also suggests the uncomfortable proximity magnifying instruments introduced between contemplation and peeping—between a nobly intellectual and a salaciously embodied form of looking.

Goldoni's telescope scene puts the resemblance between reputable contemplation and unseemly peeping on comic display. Buonafede comes to Ecclitico with a purely intellectual curiosity about the moon—a desire for some "theory" regarding its nature. At the telescope, however, his intellectual curiosity is transformed into lusty enthusiasm. As Buonafede describes what he has seen in a series of arias, the pleasure of viewing overrides any knowledge gained: "Oh, what a blessed world! Oh, what great joy! What pleasure, what delight, what fun it gives me!"[20] By exposing the similarity between natural philosopher and Peeping Tom, the scene holds investigators of nature up for ridicule.

Goldoni's scene also credits the telescope with granting access not just to the distant but also to the interior of things. While fantastic in the context of Ecclitico's extravagant claims, the idea that magnifying instruments penetrated interiors was founded in scientific practice. Hooke's characterization of microscopical observation as "peeping in at the windows," for instance, appears in his description of a water-insect with "transparent shell" that allows the observer to see inside the living creature. The ability to look within was commonly featured in treatises on microscopy, which by the 1730s were being pitched to a general audience of amateurs and connoisseurs.[21]

The rise of scientific writing for general audiences helped transform the popular culture around magnifying instruments from one of satirically minded critique to one of acceptance, even idealization. Here, too, roots are traceable to late seventeenth-century France. Bernard de Fontenelle's *Entretiens sur la pluralité des mondes* (1686) introduced a popular style of scientific writing that put the latest astronomical theories in layman's terms (or rather, laywoman's—the text was pitched especially to women), to great international success. Francesco Algarotti's *Il Newtonianismo per le dame* (1737)—also widely translated and frequently reprinted—likewise presented the discoveries of natural philosophers in dialogues between a learned gentleman and naive lady, enlivening the subject matter with conversational banter appropriate to genteel society. Popular scientific literature expanded with the work of authors who were also instrument makers, such as Benjamin Martin in England and Cosmus Conrad Cuno and Martin Frobenius Ledermüller in Germany. These authors placed particular emphasis on the telescopes and microscopes (among other devices) that they offered for sale.

The intersection of philosophical and commercial interests enabled telescopes and microscopes to transition "from experimental instrument to philosophical furniture," to use Jan Golinski's phrase for the movement of such apparatus from the specialist domain of the laboratory into middle- and upper-class households.[22] The contrast between Bordelon's frontispiece and one from Benjamin Martin's multivolume *The Young Gentleman and Lady's Philosophy* shows the resulting shift in the popular image of magnifying instruments (figure 1.2). Featuring a telescope in a well-appointed parlor between a lady and the gentleman who instructs her, the frontispiece illustrates the acceptance of magnifying instruments—and with them the desire for knowledge "beyond mankind"—into cultured life.

In popular scientific literature, cautionary tales about the error-prone nature of instrumental observation gave way to images of easy use. Still, however, this ease came with interpretive guidance. Thus, instead of exchanges like that between the clergyman and amorous lady, one finds episodes like the one described in Benjamin Martin's *The Young Gentleman and Lady's Philosophy* (1782), when a learned gentleman shows his sister the microscopic structure of mouse hair. Upon viewing the hairs, she wonders what the purpose of such "hidden Beauties" can be. Her brother replies by adapting a line from Pope's *An Essay on Man*: "*whatever is, is right*, whether you understand it or not."[23] In a striking act of reinterpretation, Martin deploys the line not to caution against investigation into realms "beyond mankind" but rather to legitimate it: phenomena observed through a magnifying instrument may exceed human comprehension, yet one knows more about nature for having seen them. Other authors offered more explicitly theological versions of the same stance: Cuno's treatise on microscopy, for example, described the observation of minute structures as ending in wonder at God's handiwork.[24] Such views presumed the reliability of magnifying instruments—that one indeed perceived through them "what is," or God's Creation. They also cultivated an interpretive stance suited to sensory extension: one need not seek to understand; instead, phenomena acquired significance through the instrumental act of uncovering them—by virtue of being hitherto hidden, and now revealed.

The ease of seeing through a magnifying instrument (which benefited from eighteen-century improvements in lens craftsmanship) was especially important to the cultivation of female audiences for natural philoso-

FIGURE 1.2. Benjamin Martin, *The Young Gentleman and Lady's Philosophy*, volume 2, frontispiece (London: W. Owen, 1772). The Bancroft Library, University of California-Berkeley, QB42.M3.

phy. However, attitudes toward female involvement in the field—even as consumers—varied regionally. While French, English, and Italian authors eagerly catered to women, Germans tended to exclude them. Immanuel Kant, for example, ridiculed Fontenelle's and Algarotti's treatises, explaining that natural philosophy was an inappropriate pastime for ladies, "deep reflection and a long drawn out consideration" being ill suited to the fair sex.[25] In seeming agreement, German popularizers of science eschewed mixed-company dialogues in favor of monologic forms of reportage. The regional difference in attitude is also reflected in the accompanying iconography. Compare, for example, the Englishman Martin's frontispiece (figure 1.2) with an image from Martin Frobenius Ledermüller's *Physicalische Beobachtungen derer Saamenthiergens* (Nuremberg: G. P. Monath, 1756): the latter shows a lone gentleman looking through his microscope at a table, his surroundings otherwise excluded from view (figure 1.3). While literature that would have magnifying instruments be entertaining and sociable

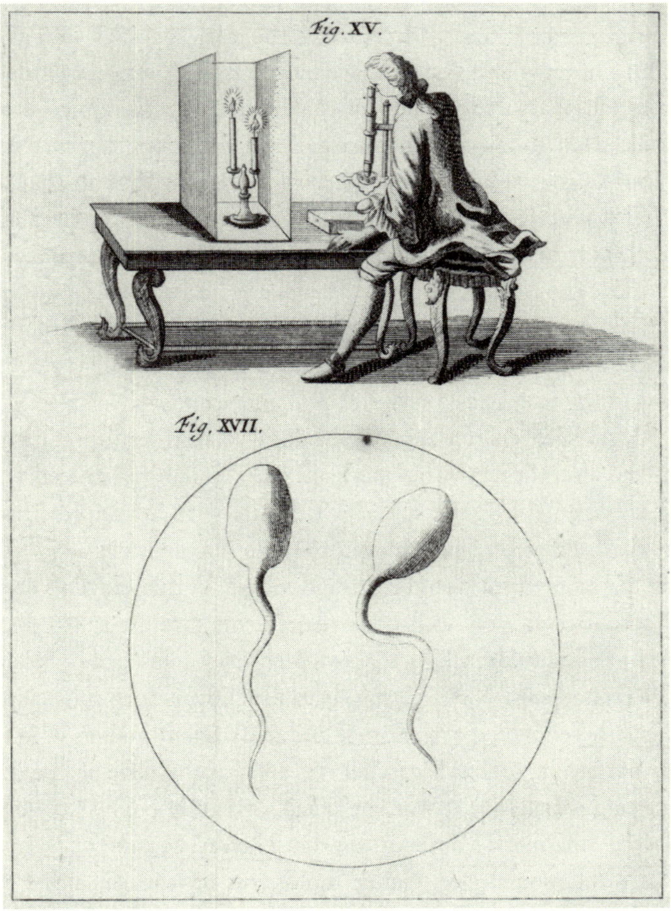

FIGURE 1.3. Martin Frobenius Ledermüller, *Physicalische Beobachtungen derer Saamenthiergens* (Nuremberg: G. P. Monath, 1756). The Bancroft Library, University of California–Berkeley, QP255.L47.

circulated internationally, their association with solitary contemplation prevailed in Germany, with consequences for the uses imagined for them and the analogies forged.

Music for Tele-Vision

Haydn set *Il mondo della luna* in 1777 for the marriage of Count Nicolaus, second son of Prince Nicolaus Esterházy, to Maria Anna Weissenwolf. Per standard eighteenth-century practice, the libretto was revised to suit local conditions, and the occasion of the courtly wedding informed the changes made. The text Haydn employed was Goldoni's up to act 2, scene 14, line 2, followed by the revised version Astarita had set in 1775, and concluding with an original act 3 finale.[26] In place of the critical tone found at the end of Goldoni's libretto and Astarita's version, Haydn's text concludes with a celebratory chorus suited to the festive occasion: "Let's rejoice, friends, at this fortune, that today comes to earth from the moon. Let us stay away from whims so that things will be better for us, now that all that we want and all that we desire goes well and in good order." Pierpolo Polzonetti has argued that Haydn's act 3 finale reflects the need of the Esterházy court to project a more progressive image in the face of shifting social and political winds. As he points out, the finale espouses key values of the American Enlightenment by figuring the moon as a new world that brings the characters "*happiness* and *prosperity* as a result of *skill* and *work*."[27]

Not only political exigencies, but also local scientific culture likely played a part in the alterations to the opera. Elaine Sisman has traced the longstanding engagement of the Esterházy family with astronomy, from Paul Esterházy's observation of a comet in 1680, to Nicolaus II's acquisition of a reflecting telescope from London around 1800.[28] While Haydn's own astronomical activities are less well documented, his interest is attested by his 1792 visit to William Herschel's massive forty-foot telescope in Slough, England. (Herschel, it is worth noting, was a Hanover-born musician turned internationally renowned astronomer and instrument maker, who believed there was life on the moon; Herschel was away at the time of Haydn's visit, however, suggesting that it was specifically the telescope that he went to see.[29]) While Polzonetti's observation that Haydn's version of *Il mondo della luna* "retains the mocking of science" holds true for the text alone, Haydn's musical treatment of the text considerably softens, even undermines, the mockery.[30] Such is most clearly the effect of the intermezzos Haydn added to the observatory scene at the start of the opera, which with the help of their scoring for muted violins work to draw the audience into the telescopic observation of a lunar world.

Although today muted violins are strongly associated with Romantic orchestrational effects of ethereality or otherworldliness, in the eighteenth century their uses were often more mundane. When German pedagogues such as Leopold Mozart and Johannes Quantz discussed the violin mute, they wrote of its ability to express certain objects and sentiments "better" or "more vividly."[31] By the time of their treatises in the mid-eighteenth century, composers had homed in on a number of phenomena for which mutes were appropriate. These included incorporeal phenomena such as dreams or ghosts; and scenes of peaceful nature, particularly when involving a babbling stream or shady grove. Typically, a text cues these topical motivations for the use of mutes, and the topics are further delineated by distinctive musical features (such as horn calls for pastoral numbers, tremolos for ghosts, and repeated undulating figures for dreams).

Throughout the 1770s and 1780s, Haydn regularly scored symphonic slow movements and certain types of operatic numbers for muted violins, his practice being typical of the period.[32] Aside from conventionalized uses, however, composers sometimes employed mutes to suit more unique circumstances. The "harmonious march" from Mozart's *Idomeneo*, discussed in the introduction, provides an example of such specialized use, where muted tone is coordinated with other musical features and dramatic context for more vivid effect: in this case, of hearing the march at a (gradually shrinking) distance.

Like Mozart's "Marcia," Haydn's intermezzos represent a special case, where the significance of muted tone is not established by convention but rather emerges from its combination with other musical features and the dramatic context. Here, too, the combination of factors suggests identifying muted tone with a change in auditory perspective—not just with hearing music at a distance, however, but with observation at a distance by means of sense-extending technology. Muted tone provides the acoustic analogue to telescopic vision.

It is useful at this point to turn to another, near-contemporary instance of the combination of muted tone with tele-vision, for it suggests that while occasions for the association were relatively rare on the operatic stage, it was nonetheless one that composers and listeners in the 1770s were ready to make. The instance comes from Grétry's *Zémire et Azor*, an *opéra comique* first performed at the Comédie-Italienne in Paris in 1771. The opera remained popular throughout Europe into the nineteenth century; at Eszterháza, Haydn directed a performance in Italian translation in 1782.[33] If Haydn knew the opera prior to composing *Il mondo della luna*, however, the evidence does not survive. Most likely, his telescopic intermezzos reflect not Grétry's influence but a parallel response to the staging of tele-vision.

In a celebrated scene, Azor (the beast in this retelling of "Beauty and the Beast") allows Zémire (the beauty) to see her distant family by means of a magic picture (figure 1.4). The same device is found in librettist Jean François Marmontel's main source for the story, *La belle et la bête*, by Jeanne Marie Le Prince de Beaumont. There, however, a mirror shows Zémire's home, Zémire reads the expressions on her father and sisters' faces, and the image spontaneously disappears. For the opera, Marmontel elaborated the episode and transformed the mirror into an audiovisual display. First, Azor tells Zémire that her family will appear in a magic picture, but warns, "if you approach, everything will vanish." The picture then shows Zémire's father and two sisters; they appear to be reaching toward her and trying to speak, but the picture remains mute. After Zémire pleads with Azor to let her hear as well, the three sing a trio lamenting Zémire's absence. In the libretto, Marmontel labeled this trio "en sourdine"—muted.[34]

In the figure of Zémire, the magic-picture scene presents an exemplary model of the absorbed spectator. As Stefano Castelvecchi and Downing Thomas have argued, the scene demonstrates the new stage-spectator relationship that mid-eighteenth-century reformers advocated for the theater.[35] As part of an effort to enhance the emotional involvement of spectators in the dramatic representation, reformers sought to establish the stage as a self-contained world. This new dramatic conception entailed what has become known as the "fourth wall"—an imaginary barrier at the edge of the stage blocking playwrights and actors from addressing or otherwise acknowledging their audience. Functioning as a "fourth wall," the magic picture lets Zémire's family be seen and heard while leaving the family unaware of having observers. Zémire, through her reactions to the picture, demonstrates "the link between spectatorial exclusion and emotional involvement" theorized by reformers of the time: wholly absorbed and profoundly moved by her family's lament, Zémire models the way operagoers should experience dramatic representations (this very scene among them) that similarly deny their presence.[36]

To this scene of model spectatorship, Grétry's music has been credited with adding the effects of enchantment and distance. The repeated three-note descending figure that opens the trio has struck modern commentators as "hypnotic," suggesting a similarity between spectatorship of the

FIGURE 1.4. Magic-picture scene from Grétry's *Zémire et Azor*, engraved by François-Robert Ingouf (after Pierre-Charles Ingouf) (c. 1771). Bibliothèque Nationale de France.

magic picture and a state of enchanted sleep (figure 1.5a).[37] Grétry scored the Trio for a wind sextet of two clarinets, two horns, and two bassoons, positioned behind the stage. In this wind ensemble, scholars have recognized the supernatural and otherworldly; in its backstage location, the evocation of distance.[38] Taking stock of the dreamlike quality and sonority of the number, David Charlton sensed a preview of musical romanticism, con-

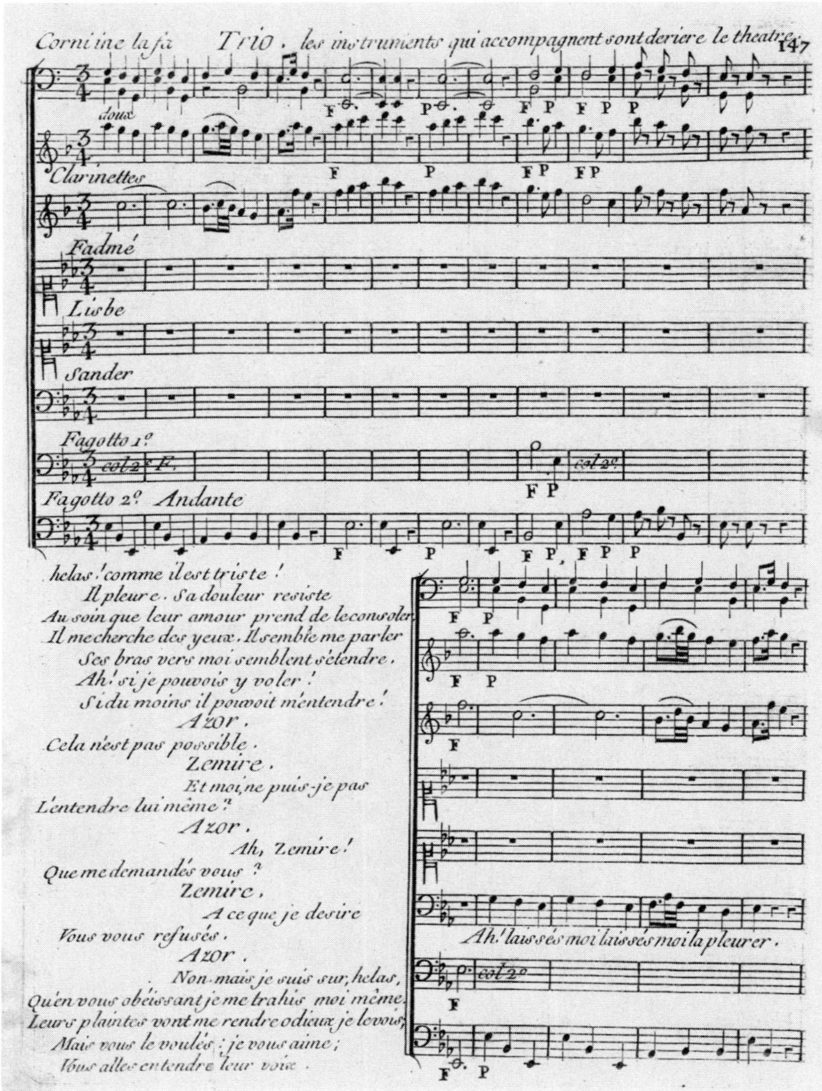

FIGURE 1.5A. Grétry, *Zémire et Azor*, magic-picture trio, beginning (Paris: Chés Houbaut, c. 1772). Jean Gray Hargrove Music Library, University of California–Berkeley, M1500.G7 Z4.

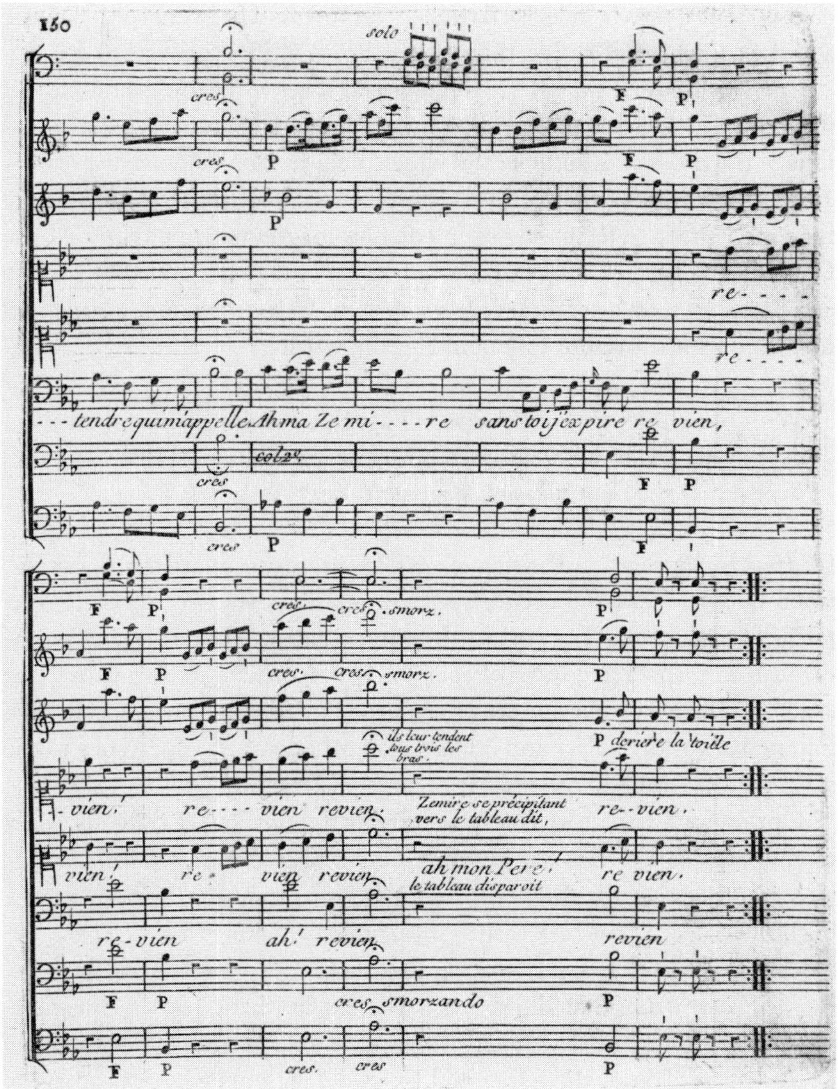

FIGURE 1.5B. Grétry, *Zémire et Azor*, magic-picture trio, end.

cluding that Grétry's "televisual trick . . . represents a union of special music and the supernatural no less than German Romantic opera was to do."[39]

The supernatural in this scene, however, takes a specific technological form; it is the technological mediation of the magic picture that the muted tone of the trio makes audible, and which gives rise to this preview of musical romanticism. Significantly, Marmontel was equally concerned to secure the visual effect of technological mediation for the magic picture. At the Comédie-Italienne, Zémire's father and sisters were surrounded by a frame and placed behind a semitransparent curtain. In his memoirs, Marmontel took credit for the semitransparent curtain, writing:

> With two ells of silver mohair to imitate a pier looking-glass, and two ells of clear and transparent gauze, I taught [the decorator] how to produce one of the most pleasing stage illusions.[40]

Gauze had appeared in Parisian stage decorations since at least the 1720s, when G. N. Servandoni used silver gauze gliding over two wheels to imitate a waterfall.[41] In the nineteenth century, gauze became a secret ingredient in atmospheric lighting effects, casting an otherworldly pall over the stage. As the mirrorlike surface of the magic picture, however, gauze both imitated a worldly object and imparted an otherworldly appearance to the stage picture; through mimesis, it became prosthesis.

Opting for a backstage ensemble, Grétry did not in fact translate Marmontel's indication "Trio en sourdine" into scoring for muted instruments (the marking *doux* for the horns indicates soft expression without damping). Yet, "muted" was how early critics, probably taking their cue from the libretto, described Grétry's music. Following the opera's premiere, the *Mercure de France* wrote that all the music was exquisite and true in its expressions of the soul, but that "one will never be able to give sufficient praise to the muted Trio of the father and his two daughters who appear in the magic picture."[42] Whereas Mozart's muted instruments made the march in *Idomeneo* sound distant, Grétry's distant (backstage) instruments made the trio sound muted. The difference between distant and muted tone acquires significance from the dramatic context: Zémire's family is not just far away, but seen and heard by means of the magic picture. The muted tone of the Trio, moreover, is tied to the visual as much as to the auditory experience of the picture: the Trio begins with instruments alone when Zémire beholds the mute image of her family, and continues with the vocal entrances once Azor has allowed sound.

In addition to coloring the sights and sounds it displays, the magic picture limits its viewers' bodily actions. Forgetting Azor's warning that the picture will vanish if she approaches, Zémire rushes forward upon an expressive highpoint of the Trio (figure 1.5b). An illustration of the scene by Pierre-Charles Ingouf captures Zémire's feeling of proximity to her father at this moment, as it seems the two, reaching toward each other, are about to touch (figure 1.4). Then, however, the picture goes dark, and Zémire is left to hear her family call "*revien*" [return] once more before their voices too disappear. By approaching the picture, Zémire does not get closer to her family but rather loses her artificial proximity. Only through the magic picture—which allows looking and listening but not touching or other bodily interaction—can she observe her distant family.

For all its scenographic and dramatic differences—an optical tube in place of the framed picture, a thrilled male rather than a sentimental female viewer, a deception rather than a revelation—Haydn's telescope scene presents a strikingly similar marriage of tele-vision and music. Haydn composed three instrumental intermezzos for the scene, one for each turn Buonafede takes at the telescope. Caryl Clark has aptly called these intermezzos "pantomimic," for they seem matched, at least in part, to the gestures Buonafede sees.[43] The first and third (identical but for the key of E-flat major and the added bassoons of the latter) feature a lyrical second phrase suited to the caressing and beseeching actions of their respective lunar scenes (example 1.1, mm. 4ff).[44] The second intermezzo lacks this lyrical material, instead featuring dotted figures and unisono passages that illustrate the violent beating in the second lunar scene (example 1.2, mm. 4ff). Though no evidence survives of how the scene was staged at Eszterháza, the pantomimic nature of the numbers suggests they were to accompany the figures seen moving inside the "moon."[45]

Haydn's intermezzos thus work to draw attention away from the total *mise en scène*, and to focus it on the "lunar world." Fixation on the "moon," indeed, seems to be the modus operandi of the first phrase of the intermezzos. Like the repeated three-note figure of Grétry's magic-picture trio, the phrase that opens Haydn's intermezzos evokes the stasis of absorbed spectatorship.[46] All three begin with the same stilted and repetitive melody over a bass line tick-tocking between scale degrees one and seven. In its opening gesture, the phrase recalls the Adagio, ma semplicemente of Haydn's Symphony no. 55, composed three years previously (example 1.3). Both employ the same melodic figure (a dotted stepwise ascent to scale

EXAMPLE 1.2. Haydn, *Il mondo della luna*, act 1, scene 3, Intermezzo no. 2

EXAMPLE 1.3. Haydn, Symphony no. 55 in E-flat major/2, mm. 1–8

degree 5, which pitch is then repeated in eighth notes), supported by an accompaniment of downbeat eighth notes separated by rests. From here, however, the intermezzos take on a static and repetitive quality not found in the symphonic slow movement. Where the first four measures of the Adagio proceed from tonic to subdominant and close with a ii-V-I cadence, the intermezzos oscillate between I and V^6; and whereas the symphonic

EXAMPLE 1.4. Haydn, *Il mondo della luna*, act 2, Sinfonia, mm. 1–5

melody falls into two halves, the first arching upward, the second proceeding downward, the intermezzo melody lacks larger shape and direction, instead reiterating the same three-note descending figure.

That this is a special kind of absorbed spectatorship is suggested by the scoring of the intermezzos. Here, too, comparison with the Adagio, ma semplicemente of Symphony no. 55 proves instructive. Whereas the symphony movement begins in two-voice texture, the intermezzo multiplies the number of voices through *divisi* scoring in the violins and double-stops in the accompaniment. With these added voices reinforcing the upper partials of the symphonic scoring, it is as if previously unheard frequencies have become newly audible. The association between the enriched sonority of *divisi* scoring and the lunar world is confirmed in the Sinfonia to act 2 of the opera, where it returns to establish the change in location from Earth to the "moon" (in reality, Ecclitico's garden) (example 1.4).

Only the telescopic intermezzos, however, combine *divisi* scoring with violins *con sordino*. Paired with the telescope, the muted violins imply that the lunar world, later encountered directly, here is perceived at a distance by means of technology. In fact, the mutes play the same trick on listeners that the telescope plays on Buonafede: they make objects that are actually close seem far away. The mutes thus do more than express or imitate more vividly; they manipulate the senses. By focusing attention on the "moon" and implying perception at a technologically mediated distance, the instrumental numbers put spectators in the same position as Buonafede: that of quietly peeping through the windows into another world. With this alteration to Goldoni's telescope scene, Haydn replaced its satire with the pleasures and promises of sense-extending technologies.

Fantasy Prosthetics

In the late eighteenth century, telescopes and microscopes—now trusted instruments of sensory extension—became increasingly common figures in philosophical and literary discourse. At one level, the discursive proliferation of magnifying instruments attests to their material proliferation, their newfound familiarity among educated people of all sorts. More significantly, it provides evidence for the impact of the instruments on habits of body and thought. Whereas Blumenberg's intellectual history of modernity identified magnifying instruments as contributing to the "legitimation of theoretical curiosity," the instruments also helped shape how that curiosity was felt and expressed. Their effects are especially evident in Romantic figurations of the desire to know the soul, or a person's inner truth. Thus, one of Ludwig Tieck's fictional creations remarked with characteristic longing, "Oh, if only we could invent telescopes to look deep into the firmament of the soul."[47] E. T. A. Hoffmann's fairy tale "Master Flea" (1822) features a tiny microscope that, when inserted into the eye, reveals to its wearer others' true, unspoken thoughts—these appearing as complex branching formations deep within their brain.[48] Whereas traditions such as physiognomy conceived the human body as "the soul made visible," inner truths being evident on the surface to the analytical eye, it was increasingly a prosthetic eye that represented the ideal means of acquiring knowledge, the mediation of a magnifying instrument seeming to provide a kind of immediacy.[49]

In proffering access to hidden regions of the soul or mind, magnifying instruments found a common purpose with another phenomenon of growing ubiquity in the latter half of the eighteenth century: keyboard fantasizing. Thanks to a confluence of factors, including the rise of keyboard instruments in the home, the cult of genius, and philosophical interest in the implications of spontaneous performance for the relation between body and mind, keyboard fantasizing—as this particularly free style of improvisation was called—spread in both discourse and material practice.[50] In 1774, J. F. Unger invented a Fantasy Machine (*Fantasiermaschine*) to capture automatically the spontaneous products of keyboard fantasizing, explaining that "many of the most powerful musical geniuses never handle their instrument more excitingly than when they gradually withdraw their mind [*Seelenkräfte*] from it completely, and concentrate on quite another object."[51] The "extraordinary state" of the keyboard fantasizer was elusive, easily disturbed by either his own conscious mind or the presence of others.

The best circumstances for encountering it thus belonged "to whoever by accident has the luck to serve, unnoticed, as an observer."[52]

When Immanuel Kant took up the topic of "representations that we have without being conscious of them" in his *Anthropologie in pragmatischer Hinsicht* [Anthropology from a pragmatic point of view] (published in 1798, based on lectures given as early as 1772), he turned to magnifying instruments and keyboard fantasizing. As Kant acknowledged, the very notion of having internal representations (or ideas) without being conscious of them seemed contradictory, "for how could we know that we have them if we are not conscious of them?"[53] In answer, he summoned magnifying instruments and keyboard fantasies as proof of the existence of such representations. Telescopes and microscopes, Kant explained, work not by bringing new images to the eye but rather by spreading out images already on the retina "so that we become conscious of them." He continues:

> Exactly the same holds for sensations of hearing, when a musician plays a fantasy on the organ with ten fingers and both feet and also speaks with someone standing next to him. In a few moments a host of ideas is awakened in his soul, each of which for its selection stands in need of a special judgment as to its appropriateness, since a single stroke of the finger not in accordance with the harmony would immediately be heard as discordant sound. And yet the whole turns out so well that the freely improvising musician often wishes that he would have preserved in written notation many parts of his happily performed piece, which he perhaps otherwise with all diligence and care could never hope to bring off so well.[54]

Kant's elaborate description of the scene and the process of improvising, with its fortuitous musical results arising from decisions beyond the musician's conscious control, illustrates his point: that keyboard fantasizing is a mechanism for bringing representations—otherwise inaccessible and destined to remain in obscurity—to consciousness. Magnifying instruments and keyboard fantasizing were both to be understood in these terms; each for Kant was a kind of internal prosthesis, bringing phenomena not from the outside world to the senses but from inside the body to the conscious mind. Just as certain images on the retina required a magnifying instrument to bring them into mental focus, certain musical ideas in the soul required keyboard fantasizing to become audible.

The analogy between magnifying instruments and keyboard fantasizing received literary treatment in the *Allgemeine musikalische Zeitung*, the Leipzig-based music journal founded by publishing firm Breitkopf & Härtel. Friedrich Rochlitz, the musician-turned-writer who served as the journal's first editor, published his story "Der Besuch im Irrenhause" [The visit to the madhouse] as a lead article in 1804. The narrative was fiction, but in the style of autobiographical travel writing in which madhouse visits were commonplace. Madhouses, at this time, not only contained the mentally ill but also put them on display to potentially edifying effect. The institution was considered a place for the sick in soul just as hospitals were for the ill in body, making a visit a means to learn about the soul's workings. Popular-philosophical writer Carl Friedrich Pockels in "Meine Beobachtungen im Zellischen Zucht- und Irrhause" [My observations in the prison and madhouse of Celle, 1794], for example, offered descriptions of the mad so as "to arouse the attention of readers regarding certain manifestations of the human soul."[55]

The aim to observe the soul likewise motivates the narrator of Rochlitz's story to visit a madhouse. However, whereas traditional madhouse visitors like Pockels studied the outward appearance of inmates' bodies for manifestations of the soul within, Rochlitz's narrator thinks of the microscope—and with it peeping past the body into the soul—as the ideal mode of observation. As he explains in the first paragraph of his account:

> Most living people held for me exactly the interest that the dead have for an anatomist, and it often pained me that one could not once grab a truly significant soul with a pair of forceps and bring it under the microscope.[56]

Fortuitously, Rochlitz's narrator discovers another means by which to observe the "exterior and interior history of an excellent young man": listening to him improvise at the keyboard.[57]

Upon arriving at the madhouse, the narrator hears unusual piano playing coming from another room. The pianist is Karl, a resident lunatic who seems to live entirely in music. When the narrator meets Karl, however, the improviser refuses to play for him. Unwilling to forego hearing more of Karl's wild improvisations, the narrator pretends to leave the madhouse, then stations himself in the room adjacent to Karl's. The asylum warden—abetting the narrator in his deception—asks Karl innocently, "Now, will you give me anything to hear?"[58] Believing the visitor gone, Karl resumes playing.

From the next room, behind a half-closed door, the narrator listens with rapt attention. He describes what he hears and its effects upon him: "As that quick section had strangely stimulated me and this slow one gently moved me, so was I deeply affected by the gradual reawakening of that wild inspiration and power which now gushed out in the most extraordinary Allegro, yet more impassioned."[59] Becoming absorbed in Karl's fantasizing, the narrator forgets his surroundings and—without realizing what he is doing—moves into the doorway to Karl's room. The madman sits with his back to the doorway, but a mirror hangs over the piano. When Karl notices the narrator in the mirror, he immediately stops playing and flies into a rage.

Like Hooke at his microscope and Buonafede at the telescope, Rochlitz's narrator adopts the position of a secret observer peeping into another world. Listening to Karl's improvising allows the narrator to observe not only exterior surfaces but also a living interior: in place of internal anatomical structures or private lunar affairs, he finds the unconscious workings of Karl's soul. By crossing the doorway, however, he throws into disarray the object he would study. The event recalls not only the vivisectionist but also Zémire at the magic picture, who by approaching violated the conditions of access. In terms of the analogy with the microscope, it is as if the narrator tried to get a closer look by moving the instrument out of the way, with the result that Karl's soul once again became invisible.

John Hamilton has pointed out that Rochlitz's narrator ultimately "confesses his incapacity to comprehend" what he observes.[60] Mystified by Karl's simultaneous playing and muttering, the narrator concludes:

> Whether his movement was entirely arbitrary, for example, like the rapid blinking of the eyes often observed in other impetuously emotional persons, or whether shapes of his then-heated fantasy hovered before him, to whom he believed he was really speaking in a private language and speaking understandably, whether by confusing "language" and "music"—(e.g., "music," "language of the heart," "without words," and the like), he mixed both up whenever he was inflamed, first conceptually and thereafter in execution; or what it otherwise meant: I do not know.[61]

The narrator's profusion of hypotheses, still not exhausting the possibilities, makes plain his inability to make sense of Karl's fantasizing. For listeners expecting a musical performance to address to them familiar ideas or emotions, such incomprehension would be a problem—a mark of failure on the part of composer and/or performer. For the eavesdropping narrator, however, unfamiliarity and incomprehensibility are legitimate qualities for the musical performance to possess. As a framework for listening, observing a soul as through a microscope provided an alternative to the imitation of familiar phenomena and the expression of conscious feelings: it figured music instead as revealing the hitherto unknown, allowing the narrator to marvel at what he could not understand, secure in the belief that he perceived truths normally beyond the reach of body and mind.

Beethoven's Eavesdroppers

Annette Richards has suggested that the lunatic Karl "mirrors and prefigures Beethoven . . . the great improviser, musical solitary, and constant scribbler whose own indulgence in musical fantasy, already the topic of some debate, was to provoke heated reaction with the appearance of the *Eroica* Symphony" within a year of Rochlitz's story.[62] Rochlitz's narrator also prefigures those listeners who sought out opportunities to eavesdrop on Beethoven at the keyboard. The first eavesdropping stories appeared in the 1820s, when John Russell reported that Beethoven's "horror of being any thing like exhibited" meant that hearing him play required a charade much like that in "The Visit to the Madhouse." At the social gathering Russell described, would-be listeners left the room where only Beethoven and an intimate acquaintance remained; by playing one of Beethoven's pieces with many errors, the acquaintance lured Beethoven's hands to the keyboard, at which point the composer could not resist beginning to play. As the composer "ran on during half an hour in a phantasy," those assembled in the next room watched and listened, "enraptured."[63] Yet at times, Beethoven's music remained beyond even their eavesdropping ears, for "while his eye, and the almost imperceptible motion of his fingers" showed him to be "following out the strain in his own soul through all its dying gradations," the instrument did not always respond, remaining as "dumb as the musician is deaf."[64]

In the numerous reminiscences published after Beethoven's death, stories of secretly listening to Beethoven from an adjacent room multiplied, often with reference to the composer's aversion to listeners.[65] Stories from Beethoven's earlier years, by contrast, painted him as an imperious performer, demanding quiet attention during his salon appearances. Citing an account of Beethoven halting mid-piece in 1802 when silence could not be secured (declaring, "I will not play for such swine!"), Tia DeNora has credited Beethoven with being "one of the first musicians to campaign consciously for a reform of the conditions of musical reception in Vienna," his actions communicating that "ritualistically solemn devotion to the performance was the appropriate form of audience conduct."[66] The offending party in this particular story was a count sitting in the doorway to the next room—the same liminal position from which Rochlitz's narrator upset Karl, here adopted not to observe the performance better but rather to achieve an optimal balance between music and conversation.

Like the pilgrimages to Beethoven examined by K. M. Knittel, tales of listening to Beethoven tell us less about "the composer per se" than about his admirers, and the positions they were prepared to adopt.[67] The salon and eavesdropping scenarios offer similar lessons in listener conduct: whether an admitted auditor or a secret observer, one should remain silent and still so as not to disturb the performer. Yet, the two story types also register a shift in listeners' adopted positions—a remapping of listening space. Where once one listened to Beethoven within a shared social circle, outside of which one expected to turn attention to other more proximate matters, by the 1820s being outside, listening in, was a desirable position. Listening at a distance, straining to catch the music sounding within Beethoven's soul, became favorite tropes—ones that Rochlitz's story links back to peeping through a magnifying instrument and the knowledge to be gained by doing so.

We might then ask whether Beethoven's works encourage fantasies of eavesdropping on the composer as a way of peeping into his soul. In Beethoven's late string quartets, modern commentators have sensed a quality of "inwardness"—of esotericism and self-absorption—that denies listeners' presence, and thus turns audience members into eavesdroppers.[68] Beethoven's contemporaries, however, might have detected the conditions of eavesdropping in keyboard fantasizing and muted tone—techniques that had acquired the ability to suggest not just secretly listening from an adjacent room but also extending the senses into the hidden regions of another world or the soul. Both an improvisatory keyboard style and muted tone feature in the Adagio un poco mosso of Beethoven's Fifth Piano Concerto, op. 73, where their implications take on added significance from the fact that Beethoven withdrew from public performance at the time of its composition.

Beethoven composed his Fifth Piano Concerto in 1809, seeing it to publication by Breitkopf & Härtel in February 1811. The last of his completed piano concertos, it was the only one that he did not himself premiere. It was additionally the only concerto Beethoven published with a fully notated solo part, the note "*non si fa una Cadenza, ma s'attacca subito il seguente*" in the first movement anticipating and preempting the soloist's desire to add a cadenza.[69] Though Leon Plantinga suggests that the shift from a contingent to a fixed text was "a natural result of publication" that simply came to the concerto later than to other genres in which Beethoven worked, Stephen Rumph has identified the work as part of "a campaign in defense of serious musical culture," reflecting Beethoven's concern to protect the integrity of the musical work from the insertion of merely exhibitionist displays by virtuosos.[70] Both recognize the Fifth Piano Concerto as novel within the genre for being a fully realized musical conception, impervious to the specifics of where, when, or by whom it might be performed—a design that may speak to Beethoven's realization that he would never himself perform the concerto.

The Fifth Piano Concerto also enjoys special status as a consummate realization of Beethoven's heroic style. Though its "Emperor" nickname is specious, its composition in 1809—the year Napoleon invaded Vienna—has lent a biographical lens through which to read its military topoi. The enlarged orchestral forces and powerful individuality of the soloist—asserted from the outset by tutti chords setting off cadenza-like displays by the pianist—further support the work's status as a concerto of unprecedentedly heroic proportions and dynamism. It is in these features that scholars have detected the work's lasting influence: Joseph Kerman called it the "prototype for the confrontational thrust of the nineteenth-century concerto."[71]

Lacking military topoi and heroic features, the slow movement of the concerto has been considered subordinate to the outer movements, and received substantially less attention. William Kinderman calls it "an immense parenthesis" in the overall design of the concerto, while Plantinga characterizes it as a "great unhurried introduction to the finale."[72] Discussing changes in the piano concerto from Mozart to Beethoven, Simon Keefe demonstrates an increasing emphasis on brilliance and grandeur in first movements, while mentioning only a concomitant trend toward "a more general atmosphere of intimacy" in slow movements.[73] The slow movement of the Fifth Piano Concerto indeed sustains a special atmosphere in ways its predecessors do not, but to call that atmosphere one of intimacy is to miss the production of distance and audience exclusion in which the move-

EXAMPLE 1.5. Beethoven, Piano Concerto no. 5 in E-flat major/2, mm. 1–10

ment engages, and which suggests interiority not through the mutual recognition of intimacy but through the unidirectional access of eavesdropping.

In its opening measures (example 1.5), the Adagio un poco mosso differs little from Beethoven's previous concerto slow movements. Like those of the first three piano concertos, the Triple Concerto, and the Violin Concerto, it begins softly with a conjunct melody, richly harmonized in homophonic texture. As in the Third Piano Concerto (1803) and the Triple Concerto (1804), Beethoven combined this dynamic level and melodic style with change to a remote key and reverberant or muted tone. The hymnlike theme presented by muted violins in the key of B major (distant from the concerto's E-flat-major key) thus draws from the same fund of expressive gestures and effects as these earlier slow movements. The novelty of the Adagio un poco mosso emerges, however, with the entrance of the piano soloist (example 1.6). Beginning with an octave leap up to F-sharp, the soloist immediately surpasses the range of the opening theme, abandoning

EXAMPLE 1.6. Beethoven, Piano Concerto no. 5 in E-flat major/2, mm. 11–29

its hymnlike melody for a distinctly pianistic gesture. The orchestra does not prepare this entrance with the typical dominant harmony. Instead, the last three measures of the orchestral introduction sit on the tonic, extending its duration through repeated figures in the upper strings and winds (example 1.6, mm. 13–15). After the soloist's grace-note pickup to measure 16, the strings repeat and sustain their previous chord, as if frozen in place by the pianist's appearance. That the piano solo enters when it does seems less a response to the orchestra than a fortuitous accident.

Ignoring the theme introduced by the orchestra, the pianist proceeds with independent material. The slow movement of Beethoven's Fourth Piano Concerto (1808) provides a precedent for such thematic independence between orchestra and soloist. In that movement, however, the contrast in musical ideas and emotional character heightens the sense of dialogue between orchestra and soloist. The alternation between *forte unisono* strings and languishing *molto cantabile* material played by the piano—the

EXAMPLE 1.6. (*continued*)

former gradually softening over the course of the movement—has earned the Andante con moto its comparisons to Orpheus swaying the furies.[74] Here, by contrast, the soloist adopts an improvisatory style, the right hand alternately soaring across the keyboard and building melodic sequences over the left hand's constant triplet arpeggiation. The strings neither wait silently for their turn to reply nor actively participate in the pianist's expression, but rather provide an accompaniment of sustained chords that could be played in spontaneous reaction to the pianist's slow-changing harmonies (an effect heightened in performance if the orchestra lags slightly behind the pianist in its chord changes).

When the soloist first comes to rest in measure 26, the strings recall measures 7–8 of the opening theme. But the pianist again takes no notice, entering in measure 28 with his earlier figuration, now a third higher. In Beethoven's other concerto slow movements, the interaction between individual soloist and collective orchestra positions the two parties on a shared

EXAMPLE 1.6. (*continued*)

stage. In the slow movement of the Fifth Piano Concerto, by contrast, the soloist gives the impression of being oblivious to the orchestra, and likewise to the presence of an audience—of believing himself alone.

The piano pedals, meanwhile, contribute to the sense of distance between pianist and listeners. The pianist's entrance in measure 16 is marked *pianissimo* and with dampers lifted, the pedal to be changed two measures later with the change of harmony. In his treatise on the performance of Beethoven's piano music, Beethoven's student Carl Czerny instructed that in this movement the shifting (*una corda*) pedal should also be used in passages marked *pianissimo*.[75] The resulting reverberant and hushed tone makes it seem as if one hears the piano not from within a shared space, but rather from somewhere apart. The sustained string accompaniment—with its muted violin tone—suggests the need to listen through something in order to perceive the piano's sound, adding to the sense of distance a material intermediary. These factors help position the listener as an eavesdropper, and more so, evoke the conditions of sensory extension.

Perhaps more important to the effect of sensory extension than listening "through" the muted violins to the piano, however, is their coordination with the piano solo over the course of the movement. Combined with the improvisatory style of his lines, the soloist's apparent obliviousness to his surroundings suggests inspired fantasizing. After a passage of trills, the fantasizing results in the discovery of the theme initially sounded by the orchestra (example 1.7). We might identify at this moment a shift from eavesdropping to mutual recognition between soloist, orchestra, and lis-

EXAMPLE 1.7. Beethoven, Piano Concerto no. 5 in E-flat major/2, mm. 42–51

teners. Alternatively, the discovery can be taken to suggest that at the opening of the movement, listeners heard the original source of the fantasizer's newfound idea. Czerny gave this source a worldly origin with the claim that "when Beethoven wrote this *Adagio*, the religious songs of devout pilgrims were present to his mind."[76] Given the process of keyboard fantasizing that the movement unfolds, however, the source may also be assigned an unconscious origin. In this connection, the muted tone of the violins contributes crucially, lending the impression that the theme sounded in the orchestral introduction issues from beyond the normally perceptible—from the heavens above or the unknown depths of the soul.

These factors converge to position the concerto listener as witness to Beethoven's creative process. As a number of commentators have pointed out, improvisatory passages in fixed compositions confuse the distribution of agency in the music-making, introducing a "slippage between composer, persona, and performer."[77] In Beethoven's Adagio un poco mosso, the improvisatory style helps conflate listening to a performance of Beethoven's composition with listening to Beethoven in the act of spontaneous music-making—a conflation that, with Beethoven no longer at the piano, could best be obtained as a carefully composed effect.

The soloist's state of inspiration is maintained until the moment the finale begins. In place of the silence that would have occasioned applause and audience acknowledgment, Beethoven composed a transition to the third movement. After the piano dies away through an arpeggiated passage on B major, the orchestra descends a semitone to B-flat. Having left the B-major region of the slow movement, the soloist hits upon a new idea: an ascending arpeggio in two-note groupings. He sounds this idea *pianissimo*, damper pedal raised, then suddenly launches into the finale by restating the figure, *fortissimo* and *allegro*, as the theme of the rondo-form movement (example 1.8). Set off as it is from the preceding material of the movement, the transition positions listeners as witnesses not so much to a process of discovery as to a moment in which the third movement seems to come to the pianist—to have come to Beethoven—fully formed.

With its sudden sonic fullness, the finale brings with it a change in auditory perspective: the implied listener is no longer a secret observer of realms beyond but rather a member of the public being addressed by the composer and the performers interpreting his work. After the immobilization, isolation, and rapt attention of eavesdropping, the rhythmic vitality and dance-like gestures of the rondo theme invite a return to the body in its shared social space. Alternatively, having practiced eavesdropping and glimpsed

EXAMPLE 1.8. Beethoven, Piano Concerto no. 5 in E-flat major/2, m. 79–iii, m. 7

the otherworldly origin of the third movement, one might ignore such cues to conviviality and continue to listen in a state of rapturous absorption. Like the magic-picture scene of *Zémire et Azor* modeled a new form of spectatorship for the theater, the slow movement of Beethoven's Piano Concerto no. 5 taught a new mode of listening for the concert hall.

From Prosthesis to Metaphysics

The association of a change to muted or reverberant tone with transport to another world or the evocation of a higher reality is now commonplace.[78] Plantinga, for example, has linked Beethoven's instructions for the piano pedal starting around the turn of the nineteenth century to a new orientation toward the metaphysical. Writing of the Largo of the Third Piano

EXAMPLE 1.8. (*continued*)

Concerto, he suggests that the instruction to lift the dampers reflects "a larger shift in Beethoven's thought, a new interest in musical sounds whose blurred surface forsakes clarity of immediate relationships to create a metaphor for a higher reality."[79] Like the violin mute, however, piano pedals were previously understood to serve imitative or expressive functions. Early nineteenth-century piano pedagogues discussed the use of tone-modifying devices to imitate other instruments: the imitative function of the lute pedal is self-evident; the moderator, known as the "celeste" in France, was said to imitate the glass harmonica, especially when one played softly and with dampers lifted.[80]

The Largo of Beethoven's Third Piano Concerto in fact sends strong signals for hearing the muted violins and lifted dampers within an imitative or expressive framework. The tutti sections of the movement invoke a pastoral mode familiar from Mozart's slow movements (and Beethoven's Pastoral

Symphony of five years later), drawing upon the conventional association of muted violins with peaceful nature. These passages encourage interpreting the piano solo to express emotions awakened by nature. Thus, a reviewer for the *Allgemeine musikalische Zeitung* called the movement an "attempt at a portrait of the melancholic mood of a noble soul," and credited Beethoven with bringing into play "all means that this instrument [the piano] possesses for the expression of gentle feelings."[81]

Through its simulation of keyboard fantasizing, by contrast, the Adagio un poco mosso of Beethoven's Fifth Piano Concerto aligns its muted and reverberant tone with the auditory perspective of the tele- or microscopic viewer and the eavesdropper. The sources discussed in this chapter thus suggest that otherworldly hearings of muted and reverberant tone have their origins not in metaphysical interests, but in prosthetic ones—in the analogies forged between magnifying instruments and musical techniques, between looking and listening through sense-extending technologies. This is to say that magnifying instruments form part of the "*technological a priori*" for otherworldly hearings of muted and reverberant tone: they made possible this perceptual experience of sensory extension, at the same time making imitative or expressive hearings of muted and reverberant tone less compelling.[82] As we have seen, however, magnifying instruments had to do more than simply come into existence: their uses, meanings, and effects had to be negotiated, consensus about their reliability reached, and they had to make their way into the hands and minds of people listening to and writing about music.

In the passage from Haydn's telescopic intermezzos to Beethoven's Adagio un poco mosso, we can also see how this level of material culture has fallen away. The explicit connections in opera and discourse became implicit, internalized in instrumental music, allowing only the end effect of otherworldliness to carry forward. Similar processes will be evident in the coming chapters. But they do not always entail technologies disappearing to leave the purely otherworldly. Visual-technological associations could equally fuel disapproving or dismissive reactions, and recede from cultural memory to leave their imprint on aesthetic standards for works of art—as we shall see by turning to opera as peepshow.

TWO

Opera as Peepshow

THE SETTING IS A WOODLAND, THE BACKDROP A GROTTO OF tangled shrubs with water descending over craggy rocks into a pool. A pair of nymphs chase after the satyr-like Komus, who carries something he has just purloined from Cupid: a peepbox. The theft occurs before the start of Johann Michaelis's comic-operetta libretto *Amors Guckkasten* (Cupid's Peepbox, 1772) yet can be seen on the title page to Christian Gottlob Neefe's vocal score, where Komus is shown grasping the peepbox behind the head of the sleeping god of desire while the two nymphs look on from the shrubbery (figure 2.1). When the nymphs catch up with Komus, they clamor for a peep in the box. Komus, however, demands that each give him a kiss before he will let them look inside. Being acolytes of the virgin Diana, the nymphs refuse to compromise their chastity while continuing to demand access to the peepshow (which they well know is an erotic entertainment used by Cupid to inspire desire in its viewers). Finally Cupid intervenes, and the opera concludes with the presentation of the peepshow. As Komus turns a crank on the side of the box, the nymphs take turns peeping inside. As they look, they describe what they see: scenes of the gods' amorous encounters, the third and final becoming so titillating that they devolve into laughter, and the stage curtain falls before they can reveal its dénouement.

A peepshow might seem a dubious choice for the opera stage. Unlike the telescope, which could be trained on a spectacular object, as we saw in the previous chapter, the contents of the peepshow remain concealed in a box, invisible to the opera audience. At the same time, the device appears to be all about vision: with its small circular window on a hidden interior, the peepshow is designed to arouse visual curiosity and to heighten the experience of looking. Barbara Stafford numbers the *Guckkasten* among the "'monolingual' or perceptual forms of communication" that traveled effort-

Amors Guckkasten,
eine komische Operette,
In Musik gesetzt
von
Christian Gottlob Neefe.

Leipzig,
bey Engelhardt Benjamin Schwickert, 1772.

FIGURE 2.1. Komus steals Cupid's peepbox while Arkadia and Hermione watch from behind the shrubbery. Christian Gottlob Neefe, *Amors Guckkasten* (Leipzig: Engelhardt Benjamin Schwickert, 1772), title page. Jean Gray Hargrove Music Library, University of California–Berkeley, M1503.N42 A5.

lessly across linguistic borders because they did not need to be read in order to be understood.[1] Identifying the peepshow as an early model of modern spectatorship, Jonathan Crary contrasts the mixed sensory modalities of the premodern fairground with the "privileging of vision" by those immobilized at the lens of the peepshow.[2]

Yet, *Amors Guckkasten* was one of several late eighteenth-century German operas to feature a peepshow centrally in both plot and audiovisual composition. Together with cantatas, novels, and dramas that likewise included peepshows, these operas clarify what iconography alone tends to obscure: that the peepshow involved speech, wordless vocalization, instrumental music, and a certain interplay of listening and looking. As we shall see in what follows, the itinerants who exhibited peepshows used their voices and often played a hurdy-gurdy or barrel organ, forming a characteristic part of the eighteenth-century soundscape. In addition to announcing and then narrating the spectacle, the peepshowman's vocal and instrumental performances seemed to exert an attractive force upon those in earshot, and to do so not through verbal persuasion but through sonic seduction. The peepshow thus produced a distinctive audiovisual arrangement: a condition of listening toward, or in anticipation of the act of looking.[3]

Recognizing the audiovisual nature of the peepshow in turn clarifies audiovisual dynamics at play in opera. Here, the peepbox created the conditions not only for listening in anticipation of looking, but also for listening in place of looking: its permanently hidden spectacle required the ear to satisfy the desires of the eye—required music and words to substitute or otherwise compensate for visual images. The incorporation of peepshows into operas thus enabled composers to explore how this substitution might be accomplished, their solutions contributing to an understanding of the power of music to activate the imagination so strongly as to equal sight. The roughly two-decade period (1770–1790) of friendly meetings between peepshow and opera came to an end, however, as the peepshow came to figure the lowly, trivial, merely sensory and mechanical elements from which early Romantics sought to extricate German opera. The peepshow thus left twin marks on musical culture, its audiovisual form persisting in a mode of listening toward, or in anticipation of visual revelation, even as its lowly status contributed to hardening distinctions between high art and mere entertainment.

Peepboxes and Their Showmen

So, what was the *Guckkasten*—literally, the peepbox—that operas like *Amors Guckkasten* incorporated into their design? The device had roots in the tradition of natural magic, with its instruments for producing wondrous effects.[4] By the mid-eighteenth century, peepboxes came in several varieties and sizes: some held wooden puppets, others paper pictures; some sat easily on a tabletop, others were large pieces of furniture; most employed a lens to enhance the illusion of depth; some also presented their contents reflected in a mirror. What all peepboxes had in common, however, was that they contained marvelous sights within them, and these could be seen only by peering through a small circular hole in the side of the box.

One could learn about peepboxes from popular scientific literature such as the Abbé Jean-Antoine Nollet's *Leçons de Physique Expérimentale* (1777), which connected the visual effect of depth or enlargement to explanations of the optical principles involved (figure 2.2). One could also purchase peepboxes (and prints to be viewed in them) from instrument makers or print shops for domestic amusement. The Augsburg printmaker Martin Engel-

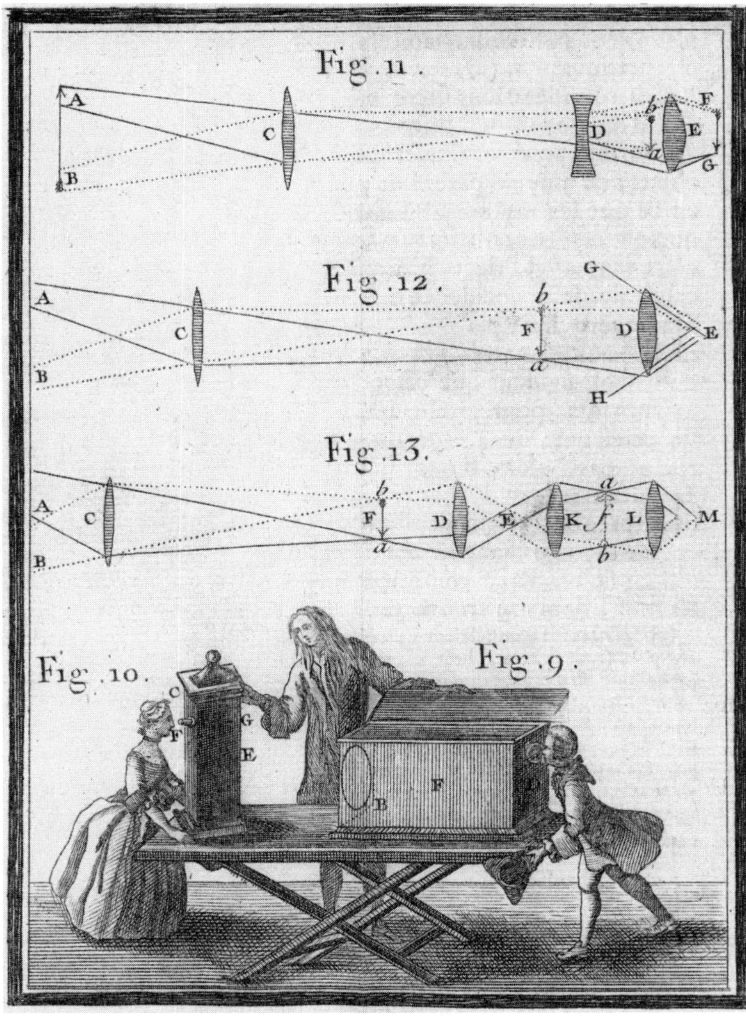

FIGURE 2.2. Abbé Nollet, *Leçons de Physique Expérimentale* (1777). Houghton Library, Harvard University, FC7.N7224.B764w v.5.

brecht (1684–1756) was an especially prolific producer of prints for peepboxes. His specialty was a form called the "perspective theater" (*Kulissentheater*) that employed layered card cutouts, each card depicting a different plane of the total scene to be viewed through a peephole with or without a magnifying lens.[5] In addition to cityscapes, biblical and historical events, and scenes from daily life, Engelbrecht offered views of the operatic stage in this form. A six-card set entitled "presentation of an opera with hovering figures" depicts a scene in front of a palace, the characters populating the

FIGURE 2.3. Ambrogio Orio (1737–1825), peepshow with hurdy-gurdy. Archives of the National Cinema Museum, Torino.

space at various depths behind a fancy proscenium (plate 1).[6] Encased in a wooden box, such prints delivered a private version of an operatic viewing experience in which the spectator sat perfectly centered before the stage, ideally positioned to appreciate the illusion of depth.

The *Guckkasten* commonly seen in paintings, engravings, porcelain figurines, and operas was not simply an optical device of the sort one could read about in treatises or purchase at shops, however. Rather, it was an entertainment presented at fairgrounds, on city streets, or in the countryside, where the optical device was inseparable from its showman (as in the Italian street scene pictured in figure 2.3, where the showman stands behind his peepbox and gesticulates theatrically). Incorporating the showman into its definition of the *Raritäten-Kasten* (as peepboxes were also called in Germany), the *Zedlerische Universallexikon* of 1741 explained it to be

> a box in which this or that old or new story is presented in miniature through puppet-work, as it may be called, fabricated for that purpose. Common people, usually Italian by birth, visit fairs in Germany with such boxes, wander the streets and through a pathetic cry, "Fine rarity! Fine mechanical-play!" [*Schöne Rarität! schöne Spielewerk*] attract to

> themselves enthusiasts who pay money to look in. Because now such things belong more to children than to adults and distinguished people, one calls things that one wants to make lowly and ridiculous "Fine rarity, fine mechanical-play" [*Schöne Rarität! schöne Spielewerk*].[7]

Though it was an instrument of popular science and upper-class amusement in certain contexts, the peepbox had a more public face as a cheap fairground or street entertainment purveyed by itinerants and on the intellectual level of children.

Peepboxes exhibited by showmen differed from their domestic counterparts in a significant respect: they were equipped with mechanisms for changing quickly between scenes, allowing the showman to efficiently cycle through a series of views. In illustrations of the "mondo nuovo," "raree show," or "boîte d'optique"—as the shows were known in Italy, England, and France, respectively—the peepshowman can often be seen working a set of strings with which he raises or lowers pictures or figures within his box.[8] In Germany, peepshowmen were more typically depicted turning a crank, their peepboxes employing one of two possible designs. The crank-operated peepbox could contain a "puppet-work," as the *Zedlerische Universallexikon* called it, consisting of marionette-like figures to be rotated into view.[9] While no known artifacts survive with this design, the peepbox in Franz Anton Maulbertsch's 1785 etching, entitled "Der Guckkastenmann" [The peepshowman], appears to contain such a puppet-work. Amid the chaos of a vulgar crowd, the showman turns a crank on the side of the peepbox; holding back a side-cloth with his other hand—and seeming to make eye contact with us, the onlooker of the scene—he allows us to see a puppet of a man holding a baby inside the box (figure 2.4). Alternatively, crank-operated peepboxes could contain a scroll of pictures that the mechanism unwound, making the pictures pass before the viewer. An example of this kind of peepbox is preserved in the private collection of media archaeologist Erkki Huhtamo: the box contains a picture-scroll system at its base and a mirror above to reflect the pictures to the peephole, where they are viewed through a magnifying lens (see plate 2).[10] The box still bears the straps by which the showman would have carried it on his back as he traveled from place to place.

The itinerant lifestyle, poverty, and ethnicity of the peepshowman contributed to his low social status, as well as to his persona. Though he might loosely be called Italian (as in the *Zedlerische Universallexikon*), he was stereotypically identified as a Savoyard; that is, as a native of Savoy,

FIGURE 2.4. Franz Anton Maulbertsch, "Der Guckkastenmann" [The peepshowman], published by Franz Xaver Stöckel (Vienna, 1785) © The Trustees of the British Museum.

the mountainous region between France and Italy. Savoyards were known throughout Europe as itinerants, members of that class of wanderers who made a living through some mix of selling, begging, and stealing. Working primarily as chimney sweeps in the sixteenth and seventeenth centuries, Savoyards journeyed to cities across Europe in the winter months to earn what little they could, returning home to farm and family in the warmer seasons. By the mid-eighteenth century, Savoyards had moved into performance, developing a number of forms of entertainment successful at attracting spectators and their money. In addition to peepshows, their performance offerings included dancing marmots, mechanical dolls, and magic lanterns (the magic-lantern show being a type of moving-image entertainment to be discussed in chapter 4). Savoyards also played musical instruments, typically a hurdy-gurdy or barrel organ. Associated with rural peasants and city beggars since the Middle Ages, the hurdy-gurdy was a semi-mechanical string instrument: it can be seen in figure 2.3, where the Savoyard standing to the side turns a crank with his left hand, which rotates a wheel that bows the strings, including at least one drone string; with his right hand, he plays the keys of a short keyboard that depress the non-drone strings to change their pitch (note: this handedness is backward for the instrument; the image was printed in reverse). The fully mechanical barrel organ joined the beggars' accoutrements in the early 1700s, and progressively displaced the hurdy-gurdy over the course of the century. Though the hurdy-gurdy and barrel organ could form stand-alone street entertainments, in the hands of Savoyards they were often paired with the peepbox or magic lantern (the partnership pictured in figure 2.3, and discussed further below and in chapter 4).

Savoyards—or at least, the people branded "Savoyards" by Europe's urban writers and artists—were the chief purveyors of moving-image entertainments for most of the eighteenth century, their migration routes constituting the distribution network for peep- and magic-lantern shows.[11] These showmen became popular objects of representation and impersonation, while in life they remained impoverished and marginalized. Much of what one would like to know about them is lost to history, their activities recorded primarily through the eyes of city-dwellers who romanticized or despised them. How they came into possession of peepboxes and magic lanterns numbers among the many unknown aspects of their lives. Their devices were handmade, without the standardization or centralized manufacturing that would characterize their counterparts in the later, industrial age. Yet, literate representations attest that Savoyards' shows tended toward a common fund of images, and were presented with a standard form of oral performance. Part of what marked the showman as a Savoyard—and made the peepshow not just a visual apparatus but an audiovisual form—was his distinctive style of hailing viewers and narrating serial pictures.

Peepshows in the Soundscape

"Schöne Rarität" [Pretty rarities] in German cities, "Voicy la Curiosita, la rareta à voir" [Here the curiosity, the rarity to see] in Paris, "O raree shoe" in London: these were the phrases that Savoyards cried to announce their peepshow to potential viewers. The cries are recorded in street-crier prints, where peepshowmen appear as one of the familiar figures found hawking his or her wares in the cities of Europe (figure 2.5). These prints, produced as broadsheets or engraved series from the mid-seventeenth through the nineteenth centuries, reflect an interest in the soundscapes of commerce,

FIGURE 2.5. Mid-eighteenth-century street-crier print, "Streetcriers of Nuremburg," depicting three Savoyard entertainers. Magic lanternist: second row from the top, first figure, cries, "Schatta Spiel an der Wand/he was curios" [Shadow-play on the wall/ Eh curious thing]. Marmot exhibitor: fourth row from the top, first figure, cries, "He schone Thiar, Schone Mürmel Thiar kane seha tantz, ah ha han" [Eh pretty animal, pretty marmotte you can see dance, ah ha han]. Peepshowman: fourth row from the top, fourth figure, cries, "Schone Rarität" [Fine rarity]. © The Trustees of the British Museum.

expressed too in musical compositions ranging from polyphonic songs and cantatas to operatic stage numbers.[12] Both capture the distinctive vocalisms with which stereotypical street-sellers broadcast their wares for sale.

The peepshowman, however, did more with his voice than the typical street-crier, for he not only advertised his show but also provided commentary upon his pictures. Widely imitated by poets, either as a vehicle for satire or in order to represent a Savoyard character, the peepshowman's commentary took the form of a descriptive list, the showman identifying the figures on view while frequently enjoining spectators to look or behold.[13] The commentary found in C. H. Weidemann's comic-opera libretto *Le Bon Vivant, oder die Leipziger Messe* (1710), in which a Savoyard presents a Raritätenkasten to two youths at the Leipzig fair, is exemplary of the showman's rhetoric of display:

So thu ich meinen Kasten auf	So I put up my box
Ein jeder legt ein Dreyer drauf	Each one pays a Dreier
Seht hier die schöne Rarität:	See here the fine rarity:
Wie dort die heil'ge Magdalena steht.	How there holy Madeline stands.
Da find die heil'gen drey Könige zart	There find the three frail holy kings
Mit weissen rohten und schartzen Bart	With white, red, and brown beard
Die schöne Cathrin auf ihrem Trohn	The pretty Catherine on her throne
Sie glänzt wie Mond und Stern schon.	She shines like the moon and stars.
Die sieben Churfürsten von Heydelberg	The seven Electors of Heidelberg
Dort sitzen sie am grünen Berg.	There they sit on the green mountain.
Die Hirsch' und Rehe jung und alt	The stag and deer young and old
Und so war Simson ehe gestallt.	And thus was Simson surrendered.
Hier ist die schöne Helena,	Here is the beautiful Helen,
Auch sitzt der König Salomon da	King Solomon sits there also
Hier ist des Königs von Babel Schmauss	Here is the King of Babel's feast
Und damit ist mein Kasten aus.	And with that my box is done.
Schöne Rarität!	Fine rarity!
Schöne Spielewerck!	Fine mechanical play![14]

The showman takes his viewers through eight unrelated scenes, the majority biblical, but with political rulers (Catherine and the Electors of Heidelberg) and a classical figure (Helen) also appearing. The focus on people suggests a puppet-work type of box; peepshows of the scroll-of-pictures type favored cityscapes over such characters. For each scene, the showman names what appears without narrative context or explanation, leaving the viewers to supply these from prior knowledge. With the last two lines, he returns to the street cry with which he advertises the show.

The peepshowman's manner of speech—his "Savoyard tone," as it came to be called—reflected his foreign ethnicity. Whether speaking English, French, or German, the Savoyard showman was typically depicted as having a heavy accent and a tendency to lapse into Italian. He also made prominent use of wordless vocalizing, record of which comes from both street-crier prints and literary sources. For example, to disparage the visual orientation of John Jakob Dusch's descriptive poem *Schilderungen aus dem Reiche der Natur und Sittenlehre durch alle Monate des Jahres*, Gotthold Ephraim Lessing likened it to a peepshow, inserting between quoted lines of the poem: "The natural Savoyard: 'Vous allés voir ce que vous allés voir! Hi! ha!'"[15] In the opera *Der Guk Kasten, oder das beste kommt zulezt*, a character calls the peepshowman an Italian "lad with his *an han han*."[16] Similar vocables appear in German criticisms of Italian opera. Telemann, for instance, complained of "the contrived ha-ha-ha, he-he-he," of stage singers, and "the Ha-he-ists' babbling trills which are considered the latest in taste."[17] Like the Italian singer, the Savoyard showman could be heard to devolve language into meaningless sounds.

The peepshow thus had two vectors of similarity with opera: in its framed visual spectacle and in its showman's voice. Both can be seen in an illustration from the c. 1715 book *Il Callotto resuscitato, oder, Neu eingerichtes Zwerchen Cabinet* [Callot revived, or the new constructed cabinet of dwarves] (figure 2.6). Inspired by Jacques Callot's "Grotesque Dwarves," the book presents a series of caricatures of dwarves in various professions and fashions. One is a peepshowman, shown carrying his box on his back and with his mouth exaggeratedly open, emitting a Savoyard-accented version of the German street cry "oh schöne Raritët": "oh sena rarite." The print identifies the peepshowman as "Nicolo Cantabella, Savoyardischer Würmschneider." A derogatory term associated with swindlers, *Wurmschneider* had the particular meaning relevant here of a "guide for strangers to a big city, who goes showing them the rarities."[18] The Italian-Savoyard surname "Cantabella," meanwhile, ironically casts the showman as one who "sings beautifully." The implicit reference to opera by way of Italian singing becomes explicit in the French and Dutch names for the showman, which were added to a version of the print published in Amsterdam: "Roger Beauchant dans l'opera des gueux" [Roger Beautiful-Song in the opera of beggars] and "Robbert Allemarkt met de Papiere Opera" [Robert All-Market with the paper opera]. These appellations capture the two ways in which the peepshow seemed operatic: the French name places the peepshowman within the many-voiced "opera" of the city street or fairground, while the Dutch likens the pictures of the peepbox to opera's visual spectacle.

The cantata *Wienerischer Tandlmarkt*, by Haydn's predecessor at Esterházy Gregor Joseph Werner, also places the peepshowman's voice and spectacle within the eighteenth-century soundscape. A "funny musical entertainment piece" (*lustige musicalische Tafel-Stücke*) published in Augsburg around 1743, the cantata is scored for four male voices (two tenors, two basses) and depicts a Viennese flea market. Each singer plays a different street-crier: a saw sharpener, a pastry seller, a mountebank, and "a Savoyard with his mechanism [*Spielwerk*]."

Werner's Savoyard proves a musical exotic among prosaic sellers of material goods. The cantata begins with a chorus that sets the market scene: each voice sings his character's distinctive street cries, as well as the cries

FIGURE 2.6. *Il Callotto resuscitato, oder, Neu eingerichtes Zwerchen Cabinet* (Augsburg: Martin Engelbrecht, c. 1715). Beinecke Rare Book and Manuscript Library, Yale University, Zg17 C14 +715.

of some additional fairground personae (radish vendor and thread seller, among others). The four voices cycle through the street cries for these various fairground traders, singing sometimes in succession, sometimes simultaneously. The Savoyard sings two separate cries: "O söner rarité, o söner Spilliwerk. Wer seh, guck in die Lok!" and "La laterna magic!" Of all the street cries, the Savoyard's are the most lyrical. The other cries are entirely syllabic, and hew to the generic figure of a monotone yell finished off with a falling third. The Savoyard's cries, by contrast, are set in the upper register, consistently the higher and more active voice in the polyphonic texture. The brief "La laterna magic!" climbs upward and downward in a melodic arch; "O söner rarité, o söner Spilliwerk" turns sinuously around the fifth scale degree. Both are set to a lilting rhythm in 12/8, while the other cries are in duple or ambiguous meters. Only the conclusion of the latter cry, "Wer seh, guck in die Lok!" falls into the monotone typical of the other voices (example 2.1).

Following the scene-setting chorus, each singer develops his character in two recitative-aria pairs. The texts of the Savoyard's recitatives reflect his heavy accent through phonetically spelled German, admixed with Italian. In the first, the Savoyard identifies himself, then solicits viewers to look at his offerings for the cheap price of a Kreuzer (equivalent to a few pennies):

Ick gom aus die Savoierland mit söner Rarité;	I come from Savoy with pretty rarities;
die Spiliwerg bell e gallant will sie vor Kreuzer seh?	Want to see the nice mechanical-play for a Kreuzer?
Da wird sie find Bell Il di Eld,	There you will find Bella Il the Eld,
Wie si mack förkt dass gansse Welt;	How she makes the whole world peer;
Le Cont de Sass sis ä su Ferd,	The count of Sass sits on his horse,
Da sittert unter si di Herd.	There under him sits the Earth.
Wer Greus ser salt vien presto qua,	Whoever has Kreuzers come here quick,
Ick spill fratant la Musica.	Until then I play music.

The arias, meanwhile, link the lilting triple meter of the Savoyard's street cries to an Italianate ethnicity. Both arias feature predominantly Italian texts and 6/8 meter. The first is entitled "Aria forlana," invoking a type of Venetian dance popular with "street people."[19] The second is a siciliana, a musical type of dubious Sicilian origin but nonetheless associated with the island's pastoral lands.[20] The numbers also confirm the association of the Italian foreigner with voice beyond speech. At the end of the first aria and second recitative, language degenerates into repeated nonsense syllables: "pe pe perle pe," in the former; "da la la ra la la ra ra," in the latter, where the recitative turns into an "arioso" to approach song.

EXAMPLE 2.1. Gregor Joseph Werner, *Wienerischer Tandlmarkt*, Chorus "Das Breinglöckl," mm. 1–24

Alone among the arias, the Savoyard's "Aria forlana" is framed diegetically, as a musical performance heard as such at the flea market. Pizzicato violins provide the accompaniment, evoking the mandolin that classically accompanied the folk dance. The preceding recitative sets up the number: "whoever has Kreuzers come here quick/Until then I play music." The phrase suggests that, setting verbal solicitations aside, music finally will draw people to the show.

EXAMPLE 2.1. (*continued*)

Such portrayals of Savoyards' manner of hailing viewers and narrating serial pictures show two kinds of relationships between auditory and visual elements in the peepshow: there are sounds (words, wordless vocalizing, and music) that advertise images, making spectators want to see them; and there are sounds (primarily speech) that accompany images, guiding viewers as they look. These audiovisual relationships were both exploited and altered when peepshows were translated to opera.

EXAMPLE 2.1. (*continued*)

EXAMPLE 2.1. (*continued*)

Peepshows in Opera

Peepshows appeared in German operas as early as 1710 as an element of passing interest, their showmen being one of many characteristic figures encountered in fairground scenes (operas with fairground peepshows include *Le Bon Vivant oder die Leipziger Messe*, the peepshowman's text from which was discussed above; and Reinhard Keiser and Johann Philipp Praetorius's *Der Hamburger Jahr-Marckt, Oder der Glückliche Betrug*, staged at the Hamburgischen Schau-Plaze in 1726). In the 1770s, librettists began to employ peepshows centrally in plot and stage action, a prominence reflected in their operas' titles: *Amors Guckkasten* [Cupid's peepbox; Leipzig, 1772]; *Der Guk Kasten, oder das beste kommt zulezt* [The peepbox, or the best comes last; Breslau, 1782]; *Das schöne Milchmädchen, oder Der Guckkasten* [The pretty milkmaid, or the peepbox; Vienna, 1796]. After the 1790s, peepshows largely disappeared from opera, though they continued to feature in spoken dramas and ballets.[21] The topicality of peepshows can partially account for their fleeting star turn on the late eighteenth-century operatic stage; the period was also rich in paintings, porcelains, and engravings depicting peepshows. But topicality alone cannot account for the ways in which librettists and composers integrated peepshows—both their oral performance and hidden spectacle—with operatic styles and forms. Equally important were the aspirations and conditions of German opera, which at this time were taking new shape.

Efforts to establish German opera against the reigning Italian opera—criticized, as by Telemann above, for its empty virtuosity, artificiality, and extravagance—unfolded fitfully over the eighteenth century. Around 1770, however, German opera entered a new phase. It was then that composer Johann Adam Hiller and librettist Christian Felix Weisse introduced a form of opera that contemporaries and later historians alike considered new—a form that, while derived from French and Italian models, was tailored to the German language and people. These operas alternated spoken dialogue with simple songs accessible to untrained voices, making them suitable both for the actors who performed German opera and for a broad public. The songs were also designed to be detachable from their dramatic context (that context typically involving rustic settings and sentimental love stories), lending them further to performance by amateurs outside the theater. As Weisse wrote, his operas "presented a pleasant tale consisting mostly of rural persons in whose mouths a small, light song sounded quite natural. These songs had melodies that were so easily grasped and so easily sung that the public swiftly retained them; they were repeated and provided entertainment in social circles."[22] Weisse's and Hiller's was a vision of opera appealing across social classes, a unifying force for a fragmented German people.

One of the librettists working in Hiller and Weisse's orbit was Johann Benjamin Michaelis, the author of *Amors Guckkasten.* Michaelis began to write *Amors Guckkasten* in 1770 while in Leipzig, publishing the libretto in 1772 after a short stint in Hamburg as poet for the Seyler theatrical company. As he explained in his preface to the publication, the idea for the operetta came from a poem of the same name he had written in 1769, while the operetta itself was a farce that he hoped would make people "laugh wholeheartedly."[23] Michaelis's emphasis on low comedy and farce set him apart from Weisse, who largely excluded such elements from his rustic, sentimental opera librettos. But as an operetta and farce, Michaelis maintained, *Amors Guckkasten* served an important purpose: such small comic pieces, coming after the main theatrical fare, drew audiences to the theater without interrupting the drama, and were the "best means to merge the liking of the crowd with the taste of the serious connoisseur."[24] A peepshow, in other words, was perfect material for a comic opera with broad appeal.

Two composers took up Michaelis's libretto, both young students of Hiller. Christian Gottlob Neefe's setting was performed in Leipzig and Königsberg in 1772.[25] Johann Friedrich Reichardt's, composed around the same time, was never staged. Both settings, however, were published in arrangements for voices and keyboard instrument (hereafter referred to as vocal scores), and it is likely that the composers had such extra-theatrical life for their music in view from the start. Hiller had discovered a sizable market for easy-to-play vocal scores of his operas, the production of which furthered his and Weisse's goal of disseminating their songs beyond the theater. Though Neefe and Reichardt could not count on the same theatrical success as Hiller to fuel sales, their vocal scores could yet find a ready audience of opera enthusiasts, keyboardists, and singers eager to make music at home.

The rise of vocal scores in the 1770s made opera available to new modes of performance and engagement. Thomas Christensen has argued that the identity of the genre transformed as a result. The circulation of operas in vocal score, according to Christensen, "shifts the aesthetic focus away from the visual to the aural, and implicitly alters the identity of the art work itself."[26] Whereas opera had primarily been considered a literary and spectacular genre, its translation from public to domestic space by means of vocal scores helped it become first and foremost a musical genre. *Amors Guckkasten*, however, suggests that the visual did not drop out of operas in vocal score so straightforwardly as an analysis of the medium alone would imply. Rather than recede in importance, the visual could be recast or sublimated into the aural dimensions of word and music. *Amors Guckkasten* makes seeing a central theme: Cupid's peepbox drives the plot, the desire of nymphs Arkadia and Hermione to look inside providing the main source of dramatic tension and action. As a consequence of the peepbox's form, however, the ultimate satisfaction of visual desire must be routed through the aural. The peepshow, being premised on the denial of access to a visual spectacle, was perfectly suited to shuttle between public stage and domestic salon: in both spaces, the visual spectacle of the peepbox can be experienced only by listening with one's ears and "seeing" in the mind.

In this light, we may ask to what extent peepshows warped the style, form, and social spaces of the operas in which they appeared. Examining three peepshow-themed operas for which music survives—the two settings of *Amors Guckkasten*, plus Joseph Wölfl's *Das schöne Milchmädchen*—shows a spectrum of accommodation between the two forms. Reichardt hewed most closely to the Weisse-Hiller conception of opera, writing simple songs that could be detached from their dramatic context and sung outside the theater. Neefe took a more experimental approach to the peepshow, sometimes abandoning conventional song styles to imitate its sounds, other times devising new ways to make its hidden spectacle visible to opera spectators.

Joseph Wölfl is best known today for his piano duel with Beethoven, but the virtuoso pianist was also a successful composer of operas. His one-act comic operetta *Das schöne Milchmädchen, oder Der Guckkasten*, on a libretto by Joseph Richter, was performed at the Vienna K. K. Hof-Theater in 1797 (where Weisse-Hiller's model was a more remote influence). The opera takes place in the contemporary "real" world of the Austrian countryside rather than in a mythological-pastoral realm, and its tenor is one of rustic sentimentality rather than farce. In contrast to Reichardt's and Neefe's *Amors Guckkasten* too, Wölfl's opera was not destined for domestic performance in vocal score. Although in 1799 Breitkopf und Härtel announced its intention to print the scores for Wölfl's "much acclaimed" operas *Das Milchmädchen* and *Der Kopf ohne Mann*, the plan did not come to fruition.[27] A publication entitled *Gesänge zu der Oper: Das schöne Milchmädchen, oder: Der Guckkasten* did appear in Vienna in 1800, but it contained only the texts of the opera's songs. These would seem precisely the conditions under which opera is more literary and spectacular than musical. Yet, Wölfl goes furthest toward recreating the audiovisuality of the peepshow, in the process foregrounding the musical-spectacular nature of opera, and an essential dynamic of listening toward looking.

Hidden Spectacles, or, "To Give the Ear Eyes"

The presentation of the peepshow marks the culminating event of *Amors Guckkasten*. After various incidents complicating Arkadia and Hermione's efforts to view the peepshow, the nymphs finally achieve their goal in the finale. By this point, Cupid has discovered Komus's theft of the peepbox but has also fallen in love with Psyche. Too happily in love to be angry about the stolen peepbox, he declares that everyone shall get a turn viewing the show. As his peepbox has only one glass, only one spectator can look in at a time while the others wait their turn.

Thomas Bauman has called Michaelis's finale "a considerable advance on the formulaic Rundgesang," in which each principal character sings a stanza of a strophic song and all echo the last line of each stanza in chorus.[28] Instead, Michaelis's finale unfolds as an alternation between the peepshowman (Komus), who announces the scene to be shown in the peepbox, and the spectator, who describes what she sees as she looks in. As each scene approaches its end, the spectator falters in her description, prompting the bystanders to ask in chorus what happens next before vying for their own turn at the peepbox.

The opera spectators, meanwhile, are left perennially without a turn at the peepbox. As numerous prints, paintings, and porcelains attest, the *mise-en-scène* of peepshow spectatorship itself held substantial interest. But there is also reason to think that Michaelis crafted the finale to encourage "seeing" the peepshow vicariously through its viewers' eyes. Not only is each scene described twice, once by Komus and once by the viewer; the onstage bystanders also perform their absorption in the hidden sights. Table 2.1 provides the text of the finale, where Komus exhibits the peepshow to Arkadia, then to Hermione, then as Arkadia, Hermione, and Psyche take turns. As becomes clear when they echo Arkadia's "and he, and he," and subsequently Hermione's "and they, and they," the bystanders hang on the spectator's every word, through which they follow the scene within the box. The viewers' descriptions of the sights in the peepbox enable fellow characters—and by extension, audience members—to envision what they cannot see with their own eyes; the bystanders in the opera model the listening-based ex-

TABLE 2.1. Text of *Amors Guckkasten*, Finale

ARKADIA UND HERMIONE Hurtig laß mich sehen.	ARKADIA AND HERMIONE Hurry, let me see.
ARKADIA Mich! (*guckt in den Kasten*)	ARKADIA Me! (*looks in the box*)
HERMIONE Mich! (*folgt ihr nach*)	HERMIONE Me! (*following after her*)
BEYDE (*indem sie einander ansehen*) Ha!	BOTH (*looking at each other*) Ha!
AMOR (*zum Komus*) Willst Du weiter drehen?	CUPID (*to Komus*) Do you want to keep cranking?
KOMUS Meinethalben.	KOMUS My pleasure.
ALLE Ja.	ALL Yes.
KOMUS Seht her, seht alle her! Wie Thetis sich am Meer im Arm des Peleus sträubt, Und Feuer wird, und Wolf, und Bär. Und Thetis ist, und bleibt.	KOMUS See here, everyone, see here! How, by the sea, Thetis Struggles in the arms of Peleus And turns into fire, and wolf, and bear And becomes Thetis again, and so remains.
ARKADIA (*indem sie hineinsieht und Komus dreht*) Da seh ich die Grotte! Da kommt er, der Mann! Die Göttin, aus Spotte, Lässt selben heran. Itzt wird sie zur Flamme— Itzt wird sie zum Lamme— wird sie zum Tyger—zum Bär— Itzt zischt sie, wie Schlangen— Nun ist—sie Mädchen—wie vorher— Und er—und er—	ARKADIA (*she looks in and Komus cranks*) There I see the grotto! There he comes, the man! The goddess, out of mockery, Remains where she is. Now she turns into a flame— Now she turns into a lamb— Now she turns into a tiger—a bear— Now she hisses like a snake— Now she—is a maiden—like before— And he—and he—
ALLE Und er, und er?	ALL And he, and he?
ARKADIA Ach, freylich hat er sie gefangen!	ARKADIA Oh, sure enough, he has captured her!
HERMIONE UND PSYCHE Hurtig laß mich sehen!	HERMIONE AND PSYCHE Hurry, let me see!

TABLE 2.1. *(continued)*

HERMIONE Mich! (*guckt in den Kasten*)	HERMIONE Me! (*looks in the box*)
PSYCHE Mich! (*folgt ihr nach*)	PSYCHE Me! (*following after her*)
BEYDE (*indem sie einander ansehen*) Ha!	BOTH (*looking at each other*) Ha!
KOMUS Soll ich weiter drehen?	KOMUS Should I keep cranking?
DIE BEYDEN HIRTEN Das versteht sich.	THE TWO SHEPHERDS I'd think so.
ALLE Ja!	ALL Yes!
KOMUS Seht her, seht alle her! Wie Vater Jupiter, als weisser Stier, aus List. Europen, übers wilde Meer Bis Kreta trägt, und küsst.	KOMUS See here, everyone, see here! How father Jupiter, In disguise as a white bull, He carries Europa over the wild sea To Crete and kisses her.
HERMIONE (*indem sie hineinsieht*) Ein niedliches Mädchen! Der freundliche Stier! Wie liebt ihn das Mädchen! Wie spielt er mit ihr! Jetzt will er sie tragen— Das wollt ich nicht wagen!— Ihr Götter! Da springt er—ins Meer!— Nun ringt sie die Hände Ey seht—da wird er—Jupiter Und der—und der—	HERMIONE (*looking in*) A pretty maiden! The friendly bull! How the maiden loves him! How he plays with her! Now he wants to carry her— This I wouldn't want to risk!— Ye gods! There he springs—in the sea!— Now she wrings her hands Ay, look—there he comes—Jupiter And he—and he—
ALLE Und der? und der?	ALL And he? and he?
HERMIONE Der macht dem Händeringen Ende!	HERMIONE He makes the hand-wringing end!
ARKADIA UND PSYCHE Hurtig laß mich sehen!	ARKADIA AND PSYCHE Hurry, let me see!
ARKADIA Mich! (*guckt in den Kasten*)	ARKADIA Me! (*looks in the box*)

TABLE 2.1. (*continued*)

PSYCHE Mich! (*folgt ihr nach*)	PSYCHE Me! (*following after her*)
BEYDE (*indem sie einander ansehen*) Ha!	BOTH (*looking at each other*) Ha!
KOMUS Soll ich weiter drehen?	KOMUS Should I keep cranking?
DIE BEYDEN HIRTEN Das versteht sich.	THE TWO SHEPHERDS I'd think so.
ALLE Ja!	ALL Yes!
KOMUS Seht her, seht alle her! Wie Vater Jupiter Sich, als ein weisser Schwan, An Leden schmiegt, wie nimmermehr Ein Schwan sich schmiegen kann.	KOMUS See here, everyone, see here! How father Jupiter, As a white swan, Nestles up against Leda, as only A swan can nestle.
ARKADIA (*indem sie hineinseht*) Ein Mägdchen im Bade! Vom weiten ein Schwan!	ARKADIA (*looking in*) A maiden in the bath! From the expanse, a swan!
HERMIONE (*die ihr nachfolgt*) Ey, wär ich im Bade, den lockt ich heran!	HERMIONE (*following after her*) Ay, were I in the bath, I'd lure him in!
ARKADIA Da kommt er gezogen.	ARKADIA There he comes, attracted.
PSYCHE (*die sich ans Guckglas drängt*) Sie scheint ihm gewogen.	PSYCHE (*she presses up to the glass*) She appears well disposed toward him.
HERMIONE (*die Psychen über die Achsel guckt*) Itzt springt er dem Mädchen aus Knie!—	HERMIONE (*looking over Psyche's shoulder*) Now he springs to the maiden on his knees!—
ARKADIA (*die beyde wegstößt*) Weg wirft er die Flügel!—	ARKADIA (*pushes both out of the way*) He throws away the wings!
HERMIONE Lass sehn! (*guckt hinein*)	HERMIONE Let me see! (*looks in*)

TABLE 2.1. (*continued*)

PSYCHE (*folgt ihr nach*) Ach, sieh doch!	PSYCHE (*following after her*) Oh, look!
ALLE DREY Hihihi!	ALL THREE Hihihi!
ARKADIA (*guckt wieder hinein*) Und sie—	ARKADIA (*looking in again*) And they—
HERMIONE (*gleichfalls*) Und sie—	HERMIONE (*likewise*) And they—
AMOR, KOMUS UND DIE BEYDEN HIRTEN Und sie?—und sie? (*Der Vorhang fällt zu und die bloßen Instrumente spielen die letzte Zeile vollends pizzicato aus.*)	CUPID, KOMUS, AND THE TWO SHEPHERDS And they?—and they? (*The curtain falls and the instruments play only the last phrase, completely pizzicato.*)

perience of the peepshow in which audience members (or consumers of the vocal score) too may participate.

In an operatic milieu where songs usually expressed feeling or delineated character, the first-person descriptions of the peepshow raise the question of what music is to do. Could music, for instance, help listeners "see" the peepshow? According to Neefe and Reichardt's teacher Hiller, it could. In a 1754 essay on musical imitation, Hiller explored the curious cross-sensory powers of music, observing:

> we often let one sense give the illusion of another or we let hearing represent things that otherwise would not be at all suitable for it. Things that should be grasped by means of an entirely different sense organ seem suddenly to have changed their nature: we believe we find them in tones, and we really do find them there, as vastly different as they otherwise are. Is this not a kind of magic?[29]

Hiller's remarks were prompted by Charles Batteux's treatise *Les beaux-arts réduits à un même principe* (1746). Borrowing Aristotle's concept of mimesis, Batteux had united the fine arts in the aim of imitating nature, and distinguished each art by the material means through which it imitates: painting through colors, dance through gestures, poetry through words, and music through sounds. Music, according to Batteux's system, could imitate "unimpassioned sounds and noises," or "animated sounds" expressive of feelings; in either case, "the sounds come from nature, and from nature alone must [the musician] take them."[30] Batteux's theory thus emphasized the single-sense nature of music: its domain was auditory phenomena.

Yet, in practice, music did not always seem so sensorially specific. Likewise responding to Batteux's theory, Rousseau also observed the mysterious ability of music to represent non-auditory phenomena and stimulate the eye. As he wrote in an entry on "imitation" in his *Dictionnaire de musique* (1768):

> Painting, which does not address the imagination but the senses—and one sense alone—only depicts visible objects. Music would seem to be similarly limited in respect to the ear, and yet it portrays everything, even those objects that are purely visible. By means of almost inconceivable powers it seems to give the ear eyes.[31]

For Rousseau, the power of music to awaken "through the medium of one of the senses responses similar to those that may be aroused by another" demonstrated that it produced stronger impressions than painting, and made it superior to the single-sense art.

Hiller was more circumspect about the power of music to make one see. For him, it was not only "a kind of magic," but also a kind of deception: listeners who believed they saw when they only heard were "deceived without becoming aware of it." Hiller wondered why listeners allowed themselves to be so deceived, and set about trying to understand the mechanisms responsible. The aria "O qual fiamme di gloria," from Hasse's *Attilio Regolo* (1749), provided an example of the cross-sensory phenomenon he had in mind: here, according to Hiller, it was clear that music—not text—was responsible for the listener's visual experience, for "one would never have heard tell of the surging blood, nor have experienced it himself, if one didn't find here the strength of a happy imitation and the hand of the master."[32]

But Hiller felt such instances of musical imitation provided no great insight into how the auditory art affected the eye, and so he turned to another, more telling object for closer analysis: the ghost. Since the ghost was unknown from nature, Hiller found, musicians instead focused on the emotional impact the object would have on the perceiver:

> The appearance of a ghost, the fluttering of a spirit departed from the body, if it really encountered or could encounter us, would be some-

> thing fearful for us. Art uses this circumstance as an aid; and as it desires to present a thing that we do not know from any experience better than from the one it presents us with, it does not seek so very carefully to express the quivering or flitting about of the ghost; it seeks to express it only fearfully, only terrifyingly. It succeeds in this stratagem. We hear and yet believe we see. The tones tremble before our ears, and we think a ghost flutters about before our eyes.[33]

Hiller thus came to the conclusion that in musical contexts, rather than objects giving rise to emotions, emotions produced objects—or rather, their perceptual illusions. Yet, even as emotional impact assumed priority in this operation, the music had to retain a degree of resemblance to the object's physical appearance: it "must still be comparable to nature and have a certain similarity to it," Hiller wrote, "otherwise the imitation would collapse."[34] Essential, ultimately, was a combination of emotional expression and imitation.

The peepshow provided an opportunity to confirm—or to challenge—such aesthetic theory with operatic practice, to explore the musical manipulation of opera spectators' visual experience. Reichardt and Neefe each musically contrast Komus's announcements of the peepshow sights with the viewers' first-person descriptions, doing so in ways that differentiate merely hearing about the sights from peeping at them. Both give Komus major-mode, duple-meter, rhythmically vigorous music, its extroverted and upbeat spirit suited to calling out what the viewer will see (as a showman would do at a fairground). When the viewers Arkadia and Hermione speak, both composers shift to a lilting triple meter, marking the viewer's transition from the public, commercial space of Komus's announcement to the pastoral, amorous space of the peepbox—and potentially transporting listeners to that new space as well.

Here, however, the similarity between Reichardt's and Neefe's finales ends, for in setting the viewers' first-person descriptions, Reichardt hewed to the priority of expressive song, where Neefe forged an alternate approach to virtual spectatorship. At the moment Arkadia begins to report, "There, I see the grotto!" Reichardt changes to triple meter and minor mode (example 2.2, m. 25). This shift, in combination with the appoggiaturas throughout Arkadia's vocal melody, yields a surge of pathos: merely looking into the peepbox—catching sight of Thetis in the grotto before any events have even unfolded—seems to fill Arkadia with sadness. She is at once visually and emotionally transported, and her song invites spectators

EXAMPLE 2.2. Friedrich Reichardt, *Amors Guckkasten*, Divertissement (Finale), mm. 10–33

to contemplate her sensitive character and state at least as much as it invites them to "see" the peepshow through her eyes.

The remaining episodes of peepshow spectatorship betray progressively lesser degrees of emotional involvement and absorption. When Hermione views the peepshow, she too switches to triple meter and sings an appoggiatura-laden melody, but remains in the same key and mode as Komus. Her less complete absorption in the peepshow is also signaled by the text: unlike Arkadia, who does nothing but describe what she sees, Hermione remarks that she wouldn't take the risk that Europa does in the scene, and uses the command form of "to see" ("Ey seht") in acknowledg-

EXAMPLE 2.2. (*continued*)

ment of her listening audience. Finally, with the last scene displayed in the peepbox, the shift to expressive song is abandoned. As Psyche, Arkadia, and Hermione each take turns viewing the scene, they sing the same duple-meter, major-mode material as Komus in his announcement. Transport, it seems, becomes impossible for viewers continually vying for a glimpse of the show; and for the opera audience, the music gives precedence to the tumultuous scene of peepshow spectatorship over the sights within the box.

Neefe's setting is less concerned with expressing the peepshow viewers' experiences—as evident, in the first instance, from the fact that he uses the same music for each viewer (Arkadia, Hermione, and the three viewers looking in turn at the final picture). To take Arkadia's turn as representative, her report, "There, I see the grotto!" begins with a shift to triple meter and modulation to the dominant (example 2.3, mm. 38–39). In contrast to Reichardt's, Neefe's melody for Arkadia lacks affective signifiers: her vocal line is entirely syllabic and without sighing appoggiaturas (example 2.3, mm. 38–78).

Key to Neefe's approach to peepshow spectatorship is not expression so much as the interaction of texted and wordless music. After Arkadia sings, "There, I see the grotto!/There he comes, the man!" instruments alone re-

EXAMPLE 2.3. Christian Gottlob Neefe, *Amors Guckkasten*, Divertissement (Finale), mm. 20–80

EXAMPLE 2.3. (*continued*)

peat the phrase, with added figuration. This pattern of alternation between vocal phrases and their instrumental repetition continues. Sometimes the figuration seems to imitate the object just described, as when Arkadia announces that "now [Thetis] turns into a tiger," and the orchestra responds with bare octaves in dotted rhythm to suggest the animal's movements (example 2.3, mm. 63–65). In most cases, however, no particular imitative re-

EXAMPLE 2.3. (*continued*)

lationship exists, a fact especially apparent when the identically decorated melody returns after other viewers describe different objects and events. Neefe's setting thus neither follows Hiller's notion of conjuring objects through emotional expression, nor relies on imitation to do so.

In addition to operating outside the terms of aesthetic theory, Neefe's instrumental interludes work against the ideal of simple, detachable song. Whereas Reichardt's peepshow viewers sing self-sufficient melody lines, Neefe's viewers have vocal parts full of gaps that do not make sense without the instrumental accompaniment. These interludes, so detrimental to

EXAMPLE 2.3. (*continued*)

the music's viability outside the opera, seem calculated to turn audiences into virtual spectators of the peepshow. The repetition of each vocal phrase without words suggests a replaying of the events described, while the added figuration encourages listeners to imagine movement. The listener can thus observe the viewer while she sings, then imagine what she has described during her silence. Whereas words tell listeners what objects and motions to see, wordless music encourages their visualization. The regular alternation between the viewer's song and wordless music invites listeners to take turns, as it were, with the viewer, to imagine looking into the peepbox with their own eyes during the instrumental interludes.

EXAMPLE 2.3. (*continued*)

In *Das schöne Milchmädchen*, Wölfl too used the strategy of alternation between sung description and instrumental interludes. Being more directly imitative of a Savoyard's peepshow, Wölfl's setting suggests that his strategy may derive from peepshowmen's exhibition practices. In this opera, the peepshow is not the goal and culmination of the action but rather a ploy toward another aim: Peter's reunion with the pretty milkmaid of the opera's title, his beloved Rose. Peter has been away, a soldier and prisoner of war for six years. He has managed to return to his country home by traveling in the guise of a Savoyard peepshowman. Addressing the peepbox in a monologue (scene 5), he credits it with his escape from captivity and successful journey. Now he asks one last service of it: to help him spy on Rose's heart—to discover whether she has taken another lover or remained true to him during his long absence.

In this service, the peepbox succeeds. Unrecognized in his Savoyard disguise, Peter exhibits to Rose, her mother Anne, her sister Lise, and her sister's fiancé Jörg an eclectic series of pictures, the subjects resembling ones found in contemporary peepshow prints: a cityscape of Vienna; a battle from the Austro-Turkish war; Delilah cutting Samson's hair; Lady Libussa (mythic founder of Prague) descending to hell; a cityscape of Venice; and finally a ship at sea in a storm. The viewers marvel at these sights, exclaiming at seeing "the entire city" or "men as if they were fully alive." At the presentation of the battle scene, however, Rose refuses to look, prompting her sister to remark, "your Peter still sticks in your head." Thus, the peepbox reveals to Peter that Rose still loves him.

The exhibition of the peepshow takes place in a quintet (for the text, see table 2.2). Similarly to the finale of *Amors Guckkasten*, the scene follows a pattern in which each new picture is described first by the showman, then by the viewers. Here, however, the showman's descriptions serve not to announce the scenes to be displayed, but rather to accompany their viewing. On first presenting each picture, Peter describes it while the viewers look in turn; continuing to look at the same picture, the viewers then exclaim in amazement and delight, naming the sights that particularly impress them, and finally, once each has had a turn, marveling in a choral refrain, "What have I not yet seen?"

In setting this number, Wölfl took the unusual step of scoring it for organ. With its motoric rhythms, arpeggiated melodies, and treble tessitura (musical features prominent in the introduction to the number; see example 2.4), the organ imitates one of the Savoyard's trademark instruments: the barrel organ. Throughout the quintet, the solo organ heralds Peter's presentation of a new picture. These eight-bar interludes, usually coming after the choral refrain—"What have I not yet seen?"—work to quiet the would-be viewers and draw them back to the show (the second instance of the refrain, followed by the organ interlude and Peter's introduction of a new picture, appears in example 2.5). The organ thus takes the place of the verbal announcements employed in *Amors Guckkasten*, providing the functional equivalent to Komus's phrase—"see here, everyone see here!"—that precedes his descriptions of the sights.

Like the line "see here, everyone see here!" the organ interludes draw the gaze while saying little about what precisely is to come. In addition to its full eight-bar phrases, the organ also alternates, at times, with Peter's description of the picture, interjecting shorter two- or four-bar phrases. Occasionally, the organ presages or echoes the sights described: a march precedes a battle scene, and a descending line in minor follows the descent of Madam Libussa into hell. More often, however, its music is generic. It does not repeat texted phrases as in Neefe's peepshow scene, foregoing this technique for stimulating non-viewers' imaginings. Instead, it follows the

TABLE 2.2. Text of *Das schöne Milchmadchen, oder Der Guckkasten*, act 1, scene 6

PETER (*mit dem Guckkasten hervorkommend, in Savojarden-Tone*) Schöne Rarität! Schöne Spielebet! Schöne Margerithe!	PETER (*coming forth with the peepbox, in Savoyard tone*) Fine rarities! Fine device! Pretty Margerithe!
LISE (*hinzuhüpfend*) Da gibts was zu sehn—was hat er denn hier guter Freund?	LISE (*skipping to him*) There is something to see—what have you here, good friend?
PETER Schöne Rarität! Aus alle Theil der Welt; komm si her schöne Kind; guck si hinein.	PETER Fine rarity! From all parts of the world. Come here, pretty child, look in.
LISE (*steht in den Kasten*) O das ist schön!—O wunderschön! Jörg, hohl geschwind die Mutter und Schwester Rose. . . .	LISE (*looks in the box*) O it is fine!—O very fine! Jörg, fetch mother and sister Rose. . . .
ANNE Wo ist denn der Guckkasten?	ANNE Where is the peepbox?
JÖRG Gleich hier vor dem Hof.	JÖRG Right here in front of the courtyard.
ROSE (*für sich im gehn*) Wenn ich nur meinen Stadtherrn nicht versäume.	ROSE (*following*) If only I don't miss my city men.
PETER (*Rosen erblickend, für sich*) Himmel! Das wahre Bild einer Rose! Nun Peter nimm deine Courage zusammen.—Schöne Rarität! Komm si her schöne Kind—O schöne Madle. (*Alle umringen nun den Guckkasten, alle wollen zugleich hineinsehen. Peter verliert kein Auge von Rosen, und drückt seine Verwunderung über ihre Schönheit aus.*)	PETER (*sees Rose*) Heavens! The true picture of Rose! Now Peter, gather your courage.—Fine rarities! Come here, fine children—O fine lasses. (*All gather around the peepbox, all wanting to see in at the same time. Peter stares continuously at Rose and expresses his amazement at her beauty.*)
LISE O seht doch liebe Mutter! Wunderschön!	LISE O see there, dear mother! Wonderful!
ANNE O lieber Gott! So was hab ich mein Lebtag nicht gesehen.	ANNE O dear God! What I haven't seen in all my days.

TABLE 2.2. *(continued)*

PETER Guck si nur immer eins nach die andere, jetzt wird ich gleich anfangen zu präsentir—belieb si wohl acht zu geben. (*Fängt die Maschine zu drehen an, guckt aber dabey immer auf Rosen hervor.*) Quintett:	PETER Look always only one after the other, now I will begin to present—keep giving your full attention. (*He begins to crank the machine, but looks always at Rose.*) Quintet:
PETER Es präsentir sich hier die Stadt, Wo si dort wohn *Empreur*; Es spazir dort auf *Esplenad* *Chapeau* mit Dame her.	PETER The city is presented here, Where the emperor lives; He strolls there on the *Esplanade* *Chapeau* with his wife.
ANNE UND ROSE O schön! O wunder schön!	ANNE AND ROSE O fine! O very fine!
LISE So lasset mich doch auch was sehn! (*guck hinein*) Die ganze Stadt, und die Allee.	LISE Let me see too! (*looks in*) The entire city, and the boulevard.
JÖRG (*guckt*) Der Stephansthurn, O Jemine!	JÖRG (*looks*) The Stephen's Cathedral, O Jiminy!
ROSE Und dorten die geputzten Herrn; Als wenn sie ganz lebendig wärn. O schön! O wunder schön!	ROSE And there the well-dressed men As if they were fully alive. O fine! O very fine!
ANNE, LISE, JÖRG, ROSE So was hab ich noch nicht gesehn.	ANNE, LISE, JÖRG, ROSE What have I not yet seen.
ZUGLEICH PETER So was hab si noch nicht gesehn.	TOGETHER WITH PETER What have they not yet seen.
PETER Belieb si nun wohl acht zu gebn! Es präsentir sich Schlacht; Wo Prinz Eugen in seinem Lebn, Viel Türk hat umgebracht.	PETER Keep giving your full attention! The battle is presented; Where Prince Eugene, in his life, Killed many Turks.
ANNE UND LISE O schön! O wunder schön!	ANNE AND LISE O fine! O very fine!
ROSE (*sieht nicht hinein, Peter drückt seine Freude aus*) Von einer Schlacht will ich nichts sehn.	ROSE (*doesn't look in, Peter expresses his joy*) I do not want to see a battle.
LISE Dir steckt dein Peter noch im Kopf.	LISE Your Peter still sticks in your head.

JÖRG (*guckt*) O schön! O wunder schön!	JÖRG (*looks*) O fine! O very fine!
ANNE, LISE, JÖRG So was hab ich noch nie gesehn.	ANNE, LISE, JÖRG What have I not yet seen.
ZUGLEICH ROSE. Von einer Schlacht will ich nichts sehn.	TOGETHER WITH ROSE. I do not want to see a battle.
ZUGLEICH PETER So was hab si noch nie gesehn.	TOGETHER WITH PETER What have they still never seen.
PETER (*in Pausen*) Nun komm si schöne Rarität, Auf andre schöne Spilebet. In *Camera obscur*, Wie si seyn *in Natur*. Der Dalila *mechant manier*, Wie si Musi Samson rasir.— Madam Libussa, wie sie schnell Marschir hinab in schwarze Höll— Dann komm si Stadt *Venezia*, Mit ihrem *Dogi*, wie si da Sich mit dem Meer verkupulir— Dann zeigt sich prächtige Revier, Wo schöne Garten paradir.—	PETER (*in pause*) Now comes the fine rarity in another pretty device. In *Camera obscur*, As they are *in Natur*. Dalila's nasty manner, As she shaves Missieu Samson.— Madam Libussa, how quickly she marches down into the black hell— Then comes the city Venice With its Doges, how it couples There with the sea.— Then appears the magnificent district Where the pretty garden parades by.—
ANNE, LISE, ROSE, JÖRG O schön! O wunder schön!	ANNE, LISE, ROSE, JÖRG O fine! O very fine!
PETER Zu letzt sich endlich präsentir, Ein Sturm, wie sich oft arrivir, Wenn man marschier auf Meer.	PETER The last to be presented, A storm, as often arrives, When one travels by sea.
LISE O Mutter, seht doch her!	LISE O mother, look there!
JÖRG Ein Schiff, als wenn's ein Stadl wär!	JÖRG A ship, as if it were a barn!
ROSE Und Blitze nach der Kreuz und Quer.	ROSE And lightning in a criss-cross.
ANNE O schön! O wunder schön!	ANNE O fine! O very fine!
ANNE, LISE, ROSE, JÖRG So was hab ich noch nicht gesehn.	ANNE, LISE, ROSE, JÖRG What have I not yet seen.
ZUGLEICH PETER So was hab si noch nicht gesehn.	TOGETHER WITH PETER What have they not yet seen.

EXAMPLE 2.4. Joseph Wölfl, *Das schöne Milchmädchen*, act 1, scene 5, Quintet, mm. 1–8

EXAMPLE 2.5. Wölfl, *Das schöne Milchmädchen*, act 1, scene 5, Quintet, mm. 107–116

EXAMPLE 2.5. (*continued*)

logic of the barrel organ, executing preprogrammed phrases like one would hear from a Savoyard. The effect is a different kind of virtual peepshow spectatorship: a transformation of the operatic space into that of a wandering showman, and of the Burgtheater's plain proscenium arch into a peephole. (There was a taste for such recursive viewing experiences: cityscapes to be viewed in peepboxes sometimes depicted a spectator at a peepshow.)

In other words, Wölfl's organ music works less to help spectators imag-

EXAMPLE 2.5. (*continued*)

ine what they cannot see on the operatic stage, and more to make them peep at what they do. The music that accompanies Peter and the viewers' singing is scored for string orchestra, and so less drastically departs from operatic norms. Sharing the duple meter and D-major tonality of the organ passages, however, the orchestra also generally continues its arpeggio-based melodies and oscillating accompaniment figures (as seen at the end of example 2.5, and in the spectators' reactions in example 2.6). Thus, in contrast

EXAMPLE 2.6. Wölfl, *Das schöne Milchmädchen*, act 1, scene 5, Quintet, mm. 25–38

to Neefe's and Reichardt's peepshow scenes, where showman and viewers had distinct music that distanced the latter from their surrounding world, Wölfl's quintet maintains the music of exhibition throughout.

There is one significant exception to this state of affairs: the passage where Rose refuses to look at the battle scene. Here, Rose's vocal line is halting, as if she has become choked up with emotion. The mode turns to minor, and a new, repeated-note figure in the violins over a descending bass line puts the music decidedly in the realm of emotional expression (example 2.7). The source of this emotion, significantly, is not the peepshow but rather Rose's inner thoughts—Peter "stuck in her head," as Lise puts it. Rather than contrasting inside and outside the peepbox, then, the quintet contrasts peepshow entertainment with inner feeling: the latter comes from a rejection of the peepbox and its visual spectacle, and is revealed musically. Wölfl's quintet thus contrasts the musical language of peepshow and opera—mechanical spectacle and emotional drama—heralding the divergence of the two media that was soon to characterize early Romantic conceptions of German opera.

EXAMPLE 2.6. (*continued*)

EXAMPLE 2.7. Wölfl, *Das schöne Milchmädchen*, act 1, scene 5, Quintet, mm. 77–88

EXAMPLE 2.7. (*continued*)

Listening Toward Looking

While scenes of peepshow exhibition balanced not-seeing and seeing, promise and delivery, another type of number focused entirely on promise: the street cry. At the start of *Amors Guckkasten*, Komus sings such a number. The text starts with a version of the typical street cry "Fine rarities," and closes with the Savoyard's trademark "hi ha" vocables:

He! Raritäten	Hey! Rarities
Lieblich zu schauen!	Lovely to see!
Püppchen und Puppen,	Dolls and puppets,
Herren und Frauen!	Lords and ladies!
Männer und Jüngferchen,	Men and lads,
Witwer und Weiber!	Widowers and women!
Götter und Götterchen,	Gods and little gods,
Täubchen und Täuber!	Little doves and doves!
Ha, hiha, trallala!	Ha, hiha, trallala![35]

Komus sings this number on the run, seeking to get safely away from the sleeping Cupid while the nymphs Arkadia and Hermione chase after him. When they finally catch up, Arkadia asks if the box Komus carries on his

EXAMPLE 2.8. Reichardt, *Amors Guckkasten*, no. 1, "He! Raritäten"

back is Cupid's peepbox, Komus confirms that it is, and the two nymphs beg to see inside.

Reichardt, as we might expect, opts for a singable melody in the folk style typical of Hiller's operas (example 2.8). There is a hint of the unrefined yodeler in the *mezzo forte* yelps in measures 6 and 8, but the overwhelming sense of the number is that of a lilting tune. The accompaniment too fits the bill of folklike simplicity: the spare two-voice texture duplicates the vocal melody in the right hand while the left provides the bass line; there is no instrumental introduction or conclusion.

Neefe, by contrast, opens the number with a substantial instrumental introduction. The keyboard arrangement bears traces of orchestral texture

EXAMPLE 2.9. Christian Gottlob Neefe, *Amors Guckkasten*, no. 1, "He! Raritäten"

in its differentiated layers (sustaining bass part, quarter-note inner voices, and more active melodic upper parts). The passage, which repeats to accompany Komus's first line, "He! Raritäten," is designed for—or to conjure—the theater: it is curtain-raiser music, complete with the registral buildup over a tonic pedal that often opened Italian opera overtures (example 2.9, mm. 1–10).

When Komus sings, he does so not in the style of a folk song but of a street cry. Whereas Reichardt subsumed the phrase "He! Raritäten" within a lilting melody, Neefe isolates and repeats its initial yell. The vocal part continues with melodic fragments separated by rests, each vocal phrase featuring the monotone typical of street cries. In one of these phrases, Ko-

EXAMPLE 2.9. (*continued*)

mus sings "schöne Raritäten," words that do not appear in Michaelis's libretto. The textual addition shows Neefe drawing on his familiarity with peepshowmen (or at least with other representations of peepshowmen) to strengthen the characterization of Komus as a Savoyard.

Neefe additionally registered Komus's Savoyard character in the accompaniment: at the end of the number, it imitates the sound of a hurdy-gurdy (example 2.9, mm. 46–48). The chords in the lower register mimic the drone

EXAMPLE 2.9. (*continued*)

of the hand-cranked instrument, while the three-note melodic figure captures the simple noodling of a street musician (similar three-note melodic figures are found in other musical imitations of the hurdy-gurdy, including in E. T. A. Hoffmann's ballet *Arlequin* of 1808, in a scene where Arlequin exhibits a peepbox).[36] The final note of the melody, strangely out of time and inconclusive, suggests the extraneous sound the strings of the hurdy-gurdy make when one ceases to crank the handle.

Dramatically, it would seem puzzling that Komus cries his "rarities, lovely to see" when he is sneaking away from the site of his theft (where Cupid lies sleeping) and being chased by the two viewers he would wish to reach. When Arkadia and Hermione finally catch up to him and the two parties begin haggling over access to the peepshow, they repeatedly remind each other to be quiet so as not to alert Diana or Cupid to their illicit activities.

The opening number makes dramatic sense, however, if we understand Komus's song to lure the nymphs to the peepbox and fuel their desire to look inside. As nymphs of the virgin Diana, Arkadia and Hermione should have no interest in an erotic show. The number thus suggests that the peepshowman's voice (and in Neefe's setting, his musical instrument) has a special force—that it arouses a desire to look, causing one to listen in anticipation of sights to behold.

That the peepshowman's "voice"—literal or instrumental—had the power to lure viewers, pique their curiosity, and awaken their desire to see is a recurring theme in the literature on peepshows. The *Zedlerische Universallexikon* of 1741 put the attractive force of the Savoyard's voice in its definition of the peepshow, noting that "through a pathetic cry 'Fine rarity! Fine mechanical-play!' [peepshowmen] attract to themselves enthusiasts who pay money to look in." The peepshowman's words, of course, played a part in this operation. "Rarity" and "curiosity" were terms connected with knowledge and desire, and while the peepshowman's puppets or pictures were hardly "rarities" of the type collected by princes and scholars and displayed in *Wunderkammern*, the promise of rarities functioned to cast the showman's box as a repository of unusual and valuable objects, and to arouse curiosity about them.[37]

Yet, tone—sound apart from words—regularly featured as crucial to the conversion of indifference into desire, its power bordering on (or revealed to be) magic. Such is the case in fictional portrayals of peepshows such as that in Friedrich Maximillian Klinger's 1791 novel *Fausts Leben, Thaten und*

Höllenfahrt. In this telling of the Faust legend, the devil disguises himself as a peepshowman in order to seduce the innocent Angelica on Faust's behalf. Stationing himself outside Angelica's window, the devil "raised his voice, and invited the people to come and see his peep-show. The populace flocked around him,—footmen and chambermaids, wives and widows, boys and graybeards. The Devil showed them all kinds of scenes, which he accompanied with pious explanations and moral sayings."[38] Like the crowd, Angelica is attracted by the devil's voice: "The beautiful Angelica now looked out of her window; and, hearing the Devil descant in so pious a tone, she felt an irresistible desire to see the wonders of his box, and to bestow alms upon the devout old showman."[39] Angelica faces dire consequences for being thus lured to the box, for the devil "unfolded to her view scenes of love, in which he led her fancy so adroitly from the spiritual to the carnal, that she was scarcely aware of the gradation."[40] By the end of the show, her heart is so inflamed and her mind so overwhelmed that she rushes to her chamber and "[sinks] senseless into the arms of Faustus"—defenseless against his sexual advances (figure 2.7).[41]

Friedrich Baron de la Motte Fouqué's story "Die Kuckkasten," published in a collection of fairy tales in 1817, likewise cast the peepshow as a trap into which sound lured the innocent. In the tale, a happy boy named Karl Grünbaum attends the local annual fair. Enjoying gingerbread and the general tumult of the street, he and his friends are drawn into a garden by the sound of merry music. There, the boys see "a strange man standing behind a peculiar little box, which looked like a little cottage full of round windows, covered with colorful curtains here and there" (a sight pictured in the story's accompanying illustration by E. T. A. Hoffmann; see plate 3).[42] The strange man, colorfully dressed and wearing distinctive red boots, calls aloud:

Heran, heran, herbei,	Step up, step up, right up,
Wer Schönes schauen will und kann!	Whoever wants to see beauties!
Ich bin dazu der rechte Mann,	For that I'm the man,
Und die Entr'e so gut als frei!	And the entré is as good as free![43]

These "alluring promises notwithstanding," no one steps up, and Karl and his friends too would have passed by, "if the foreigner had not lifted a small barrel organ to play, around which many light silver bells turned in the best measures, and repeated very delicately the ditty of the above little verse."[44] At this, Karl suddenly finds the promise of free entertainment irresistible and corrals his friends in order to take the strange man up on his offer. Pres-

FIGURE 2.7. "She saw before she could flee." Illustration from Friedrich Maximillian Klinger, *Fausts Leben, Thaten und Höllenfahrt*, in 5 vols. (St. Petersburg, 1794). The Bancroft Library, University of California–Berkeley, PT2380.K7 F72 1794.

ently, the man reveals enough windows to allow all the boys to peep into the box at once, and as the boys look at Constantinople from the distant perspective of a ship, the man explains that if the boys wish they can enter into the scene and steer the ship closer to the city. Karl, skeptical that this is possible, decides to give it a try and finds himself fully immersed in the peepbox scene, the ground shifting beneath his feet as he stands on a ship being tossed about at sea. At first, Karl believes he is still experiencing an entertainment and marvels at the completeness of illusion. Eventually, however, Karl learns that he has truly been transported to the exotic land he sees around him, where he is held captive by the showman who is in fact an evil magician.

In these stories, the tone of the voice or of a musical instrument provides the prelude to peeping—to an act of looking into another world. This audiovisual dynamic characterizes a mode of listening different from the "prosthetic" mode discussed in the previous chapter. Analogies with magnifying instruments familiarized how to listen to music as coming from regions of the heavens or soul beyond the normal reach of the senses, and to accept what seemed incomprehensible as the revelation of previously hidden phenomena, the very production of knowledge. The peepshow, by contrast, provided a framework for listening within which music, meaningless in itself, pointed to another world—and made the desire to behold that world overpowering.

Das schöne Milchmädchen has no number like Komus's "He! Raritäten"; Peter's street cries appear only as spoken text. But Wölfl found an opportunity for equivalent music—music devoted to arousing a desire to see—in a standard part of the operatic form: the overture. Where Neefe imported the curtain-raiser into the peepshowman's cry, Wölfl translated the peepshowman's cry into the overture. The main theme imitates a barrel organ. Scored for clarinets in place of the usual violins, the eight-measure theme evokes the mechanical instrument through the same features as the peepshow quintet: motoric rhythms, copious arpeggiation, diatonicism, and a repetitive antecedent-consequent phrase structure (example 2.10, mm. 5–12). The main function of the opera overture has often been characterized as capturing attention and preparing the audience for the theatrical performance. Different here is the recognition that the opera overture shares this function with the peepshowman's barrel organ music. If early eighteenth-century observers detected operatic qualities in the peepshowman's voice and spectacle, the later eighteenth century discovered a new, audiovisual convergence between opera and peepshow: both were forms in which music made one listen desirously toward looking and visual revelation.

Rereading History Through the Lens of the Peepshow

For a number of scholars, the peepshow has become a harbinger and emblem of modernity. In *Bach's Cycle, Mozart's Arrow*, Karol Berger selected Giandomenico Tiepolo's 1791 painting of a peepshow to represent the mod-

EXAMPLE 2.10. Wölfl, *Das schöne Milchmädchen*, Overture, mm. 1–12

ern, linear (as opposed to the premodern, cyclical) consciousness of time. In Berger's reading, the painting, entitled *Il mondo novo* [The new world], shows

> a thoroughly modern crowd assembled to gawk at a spectacle made possible by the newest technological medium (shortly to be featured also in *Faust II*)—a magic lantern displaying the exotic marvels of "the new world." These humans are not subject to an eternal, unchanging order. On the contrary, they are children of a unique historical moment, their gaze fixed on a dimly imagined future, a new, emerging world. Tiepolo's time is linear, progressive, oriented toward the future.[45]

EXAMPLE 2.10. (*continued*)

While this reading makes Tiepolo's painting a fitting heuristic for an emergence of modern temporal consciousness c. 1790, a reading of the painting grounded in the history of the peepshow yields different conclusions. The first correction is chronological: by 1791, the peepshow, far from being "the newest technological medium," was a well-worn staple of streets and fairgrounds (Tiepolo himself painted a very similar *Il mondo novo* c. 1765; Goethe's *Faust II* did invoke the recent technological innovation of the

"Phantasmagorie," but this involved not a peepshow but a form of magic-lantern show developed in the 1790s—a subject of chapter 5).[46] The second concerns the nature of the spectacle. Erkki Huhtamo has observed that the peepshow in Italy—the "Mondo nuovo," as Italians typically called it—was a more sophisticated entertainment than in other countries, a fact evidenced not only by the greater size and ornateness of the boxes but also by the existence of broadsheets describing their programs. Yet, in these peepshows, as in those presented throughout Europe, views followed one another in a succession that was seemingly random, "dictated by availability," Huhtamo suggests, "or perhaps by the logic of attraction that emphasized the display of curious sights over narrative continuity."[47] So, while the spectator's view was fixed straight ahead, what he or she saw in the peepshow was far from a linear narrative: rather, it was a series of pictures exemplifying Berger's premodern temporality, in which events do not unfold according to one, progressive path but instead make sense regardless of their order.

Most significantly, for the early Romantics who, by Berger's account, embraced the linear time of modernity, the peepshow represented a backward, regressive form. Offering a capsule history of opera in Germany, Carl Maria von Weber described how during the Napoleonic Wars—now two years behind—opera degenerated into a lowly, escapist entertainment:

> Depressed by the horrors of war, having become familiar with all manner of misery, one sought the most basely-tantalizing artificial pleasures in order to cheer oneself up. The theatre turned into a peepbox in which—fearfully avoiding the beautifully enriching emotional upheaval that arises out of the true enjoyment of a work of art—one allowed a series of scenes to unroll, satisfied to be tickled by trivial fun and melodies, or blinded by machine gimmickry without purpose or meaning.[48]

Emotionally exhausted by reality, people sought escape, and trivial entertainments obliged by delighting passive spectators with aimless series of scenes and melodies.

To turn into a peepshow, in this sense, was to move away from the ideal of German Romantic opera. Weber was optimistic for the future, however, for just as "Gluck's creations . . . appeared in that time in which lustful Italian musical waves had engulfed and softened all minds," a truly Romantic creation had now emerged from the war-weary land: E. T. A. Hoffmann's *Undine*, to a libretto by "Die Kuckkasten" author Fouqué. This was what the opera theater needed: "a work of art that is complete within itself, in which all parts and contributions of related and employed art forms melt into each other and thus vanish and, by becoming lost to a certain extent, create a new world."[49] According to Weber, the difference in the operatic form—the unified whole as opposed to the serial arrangement of heterogeneous elements—translated into a difference in audience behavior. "The malicious attribute too much of the work's success to the stage sets," he wrote in reference to Karl Friedrich Schinkel's magnificent designs, which included a sumptuous water palace and Gothic tomb. "However, the reviewer notes that in other pieces where this is the case, the public just waits for these spectacular moments and then leaves, whereas here they remain with constant, equal attention from beginning to end . . ."[50] The "new world" of *Undine* was not one of interchangeable parts suited to intermittent sampling but rather a continuous whole requiring engagement over its entire duration.

Twenty years before Weber's review of *Undine*, commonalities between opera and peepshow presented opportunities for incorporation, accommodation, and experimentation. Now, they spurred efforts at differentiation and divergence. The lowly entertainment provided a model against which to define the forms and aims of opera as high art. The autonomy, unity, synthesis, and continuity that Romantics prized were decidedly absent from the Savoyard's peepshow, which became—beyond a cheap street show—the epitome of a poor mixture of visual, verbal, and musical arts. The appeal of the peepshow that fueled its appearance in opera was discarded along with the key priorities of Hiller and Weisse's artistic program (universal song circulation, bridging high and low), to be replaced by the Romantic ideology of the work of art.

Yet, the divergence of high art and low entertainment was greater in theory and rhetoric than in practice. As Martin Nedbal has noted, discussing the prevalence of puppets in recent productions of Mozart's *Die Zauberflöte*, the high art of opera is commonly opposed to the naive spectacle of puppet theater, leading critics to assume that productions employ puppets merely for popular appeal, and at grave risk to the masterwork. In fact, however, "the inception and reception history of the opera has always been marked by close encounters with this type of theatrical media."[51] The observation can be made more generally of German opera and technological media of the late eighteenth century and beyond. Weber's own operatic triumph, *Der Freischütz* (1821), exemplifies the continuation of such close encounters. Anthony Newcomb has argued persuasively for the influence of moving picture entertainments—specifically the phantasmagoria—on Weber's conception of the Wolf's Glen scene. With Adorno, who described the Wolf's Glen as "made up of a succession of little pictures," he agrees that the scene marks "a distinctive intersection of burgeoning technology, instruments of popular amusement, and serious opera."[52] Important here is the qualifier "serious": the technology-amusement-opera intersection was not new; as we have seen with the peepshow, it was well established for German comic opera. German Romantic opera, however, secured its seriousness in part by turning such openly acknowledged borrowings into covert and anxious ones.

If the peepshow appears as musical modernity's Other in terms of temporal organization and cultural aspirations, it proves less at odds with another vector of modernity it has been taken to herald. Recognizing that the form of the peepbox dates from at least the seventeenth century, Jonathan Crary sees it transitioning from a minor element of early modern popular culture to an important model for modern visual culture's dominant forms, which turned "the multifaceted festival *participant* . . . into an individualized and self-regulated *spectator*." For Crary, the device figures the "psychic, perceptual, and social insularity" of the modern viewer, who ultimately internalizes the imperatives for discipline and social restraint to become silent, immobile spectators.[53]

In identifying the peepshow as a model, Crary thinks of its resemblance to other, later apparatuses that fix their viewers to their eyepieces (like the stereoscope, Kaiserpanorama, and kinetoscope). But shifting from a visual to an audiovisual conception of the peepshow allows us to consider what it holds in common with opera, and how its brief prominence on the operatic stage helps articulate a change in the art form's privileged modes of engagement. The act of looking—the routes taken by, and the effects of the gaze—has garnered substantial attention in recent opera studies, thanks in large part to the heightened awareness of the visual and scripting of the gaze that has come with "filmed" opera. This awareness has opened up new interpretations of the story of Orpheus, long considered opera's founding archetype. Thus, where Theodor Adorno claimed that "all opera is Orpheus" by virtue of a shared intervention of music in fate, and Carolyn Abbate suggests regarding the genre of opera as a response to the question of how Orpheus's head can go on singing after his death, Michal Grover-

Friedlander has argued for an essential relation between voice and gaze: just as Orpheus regains Eurydice from death through his song only to lose her again by looking at her, in opera, Grover-Friedlander argues, "the visual, or the optical is able to bring about the total collapse of whatever has been achieved by the vocal or the acoustic."[54]

The audiovisual dynamic of the peepshow invites an alternative reading of the relation between the visual and the acoustic in both Orpheus and opera. For while one may regard the gaze as undoing what has been achieved by the voice, it is also the work of music to bring about occasions for irresistible, heightened looking. Orpheus's persuasive song makes possible his fatal look; but even more immediately, listening compels him to look at what is hidden/forbidden from his sight. In Monteverdi's version of the legend, Orpheus—having been commanded by Pluto "never [to] turn his desirous eyes to see her/Since her eternal loss/Will be caused by a single glance"—leads Eurydice from the underworld, resisting the love and worry that make adhering to Pluto's prohibition so difficult.[55] But when a noise is heard behind the set, Orpheus, fearing it is the furies come to take Eurydice back, turns to look; the sound thus dissipates his resistance and prompts him to catch the longed-for yet fatal glimpse of Eurydice. In Gluck's version, Love forbids Orpheus to look at Eurydice until they are outside the Stygian caves. As he leads her by the hand, Eurydice pleads with Orpheus to look at her, taking his refusal as a sign he no longer loves her. "The more I listen, the less I resist" (Più che l'ascolto/Meno resisto), Orpheus opines; finally, he turns and looks at her. While verbal persuasion plays a large role in this case, in both operas—as in the peepshow stories discussed above—the ear is the conduit for an irresistible draw on the gaze.

Both the Orpheus legend and the peepshow thus entail an audiovisual dynamic of listening toward looking; but there is also a significant difference between the two. Common to all versions of the Orpheus legend is the turn to look, an action that roots vision firmly in the body and its environment. The look of the peepshow, by contrast, exits the viewer's environment to enter another world, where the body cannot follow. This transformation of the look *on* the operatic stage (performed also by the telescope discussed in the previous chapter) presaged a transformation of the look *at* the operatic stage. According to Weber, audiences were ready to greet *Undine* with a spectatorship of constant attention through which they entered its "new world." Peepshows provide a dubious model for prolonged silence and immobility, as their exclaiming and turn-taking viewers in opera make clear;

yet, they familiarized—in small doses—a kind of transport further cultivated in German Romantic opera.

Listening toward looking into another world—this distinct form of audiovisuality—was not confined to operas and peepshows. It also became a way of experiencing instrumental music—a framework for listening that, like the prosthetic mode of sensory extension, made sense of music's not making sense. Carl Friedrich Zelter wrote in 1809, for example,

> There are certain symphonies by Haydn that in their loose, liberal progression bring my blood into a comfortable motion and give the free parts of my body the inclination and tendency to work outwardly. At these times, my fingers become softer and longer, *my eyes wish to see something that until now no eye has ever glimpsed*, the lips open themselves, that which is within me wants to go out into the open.[56]

Mark Evan Bonds cites Zelter's remark as illustrative of a shift in the premises of musical listening from rhetorical and representational to idealist ones, the latter regarding music as a "philosophical language" communicating ineffable truths. Arguing that Zelter mixes "older (passive, physical) and newer (active, mental) modes" of listening, Bonds explains: "Zelter described the act of listening as a process in which a physical response led to a heightened mental striving for a world beyond . . . the spirit transcends the realm of the phenomenal and catches a glimpse of that which would otherwise remain inaccessible."[57]

Yet, what Bonds reads as a "mental striving" is articulated by Zelter as a physical striving by the eyes that wish to behold the hitherto unseen. When speaking of listening under the rubric of transcendence, Bonds too makes vision central. Both here and elsewhere, Bonds's sources lead him to describe music as the route to a visual encounter with a higher realm. The aim of listening, the orientation of the longing it arouses, is toward a "glimpse" of the infinite or Absolute.[58] Listening proves fundamentally scopophilic, filled with ocular desire. The centrality of vision is especially evident in the cover art for Bonds's book: from French architect Claude Nicolas Ledoux's treatise *L'Architecture considérée sous le rapport de l'art, des moeurs et de la législation* (Paris, 1804), the illustration shows a theater's tiered seats, reflected in the iris and pupil of a giant eye. The eye, here, is meant to stand as a metonymy for the mind and philosophical thought.[59] But such metonymy represents only a partial slice of the period's visuality, a limited sense of what it could mean to wish to see. Opera and peepshows—like other audiovisual forms—provided spaces of looking and listening in which flourished less abstract, more practiced forms of vision directed toward other, hitherto unseen worlds; vision stimulated by music.

We will encounter Zelter—a friend of Goethe, composer of lieder, and director of the Singakademie in Berlin—again in chapter 4, for he also wrote of experiencing Haydn's oratorio *The Creation* as a shadow-play for the inner eye. First, however, we turn to the shadow-play for the outer eyes—an apparatus that framed views not with a peephole and lens but with a screen and the boundary where darkness meets artificial light.

THREE

Shadow Media

THIS SPECTACLE, PRESENTED BY SHADOWS REALLY SPEAKING and acting in a certain order, is probably the first of its kind."[1] So read a report on a shadow-play performed in Regensburg in 1781. The show's organizers could not claim to have invented the spectacle of "shadows really speaking and acting." The name they used for the spectacle, "Chinesische Schattenspiel," acknowledged the particular form of shadow-play that had emerged in 1770s France and already found its way to other parts of Europe—a form in which silhouette puppets acted out scenes with dialogue and music.

But the spectacle was "probably the first of its kind" in another way, not for the shadow figures speaking and acting but for their use to present a previously stand-alone musical composition: Johann André's setting of Gottfried Bürger's ballad "Lenore." Whereas the genre of Chinese shadow-plays often mixed simple airs with spoken dialogue and accompanying instrumental music, the Regensburg performance staged a through-composed song that had been published with no eye to such realization. This song, as we will see later in this chapter, earned its composer André the nickname "father" of the through-composed ballad, when an early nineteenth-century critic identified it as the first in a movement that left behind the strophic approaches to ballad composition typical in the eighteenth century.

The Regensburg shadow-play thus took place near the beginning of a transitional period for the musical genre of the ballad. Though debate continued about the merits of setting ballads in simple folk song or in dramatic art-music style well into the nineteenth century, the weight of aesthetic theory and compositional practice shifted toward the latter in the decades around 1800. The reasons for this shift, which also affected German song composition more broadly, have often been pinned to literary developments: German poetry was becoming more psychologically progressive, and so "demanded" the synthesis of strophic and through-composed form that became a hallmark of the nineteenth-century lied. But such explanations fail to recognize that far from presuming static poetry, the advocates of strophic settings articulated principles that were entirely compatible with narratively and emotionally diversified texts. More than a response to changes in literature, the shift toward through-composition had to do with modes of reception and listening—with a wider transformation in the kinds of experiences available to consumers of culture.

Here, Friedrich Kittler's analysis of reading practices at the turn of the nineteenth century provides an instructive complement to changes in musical listening. According to Kittler, reading fundamentally changed as a result of new pedagogical methods that focused on the meaning rather than the sound constituents of printed words. The focus on meaning enabled the words on the page to become invisible, to provide seemingly direct stimulus to the imagination. While such stimulus was achievable as a special poetic effect under the old methods, the new pedagogical methods made it an automatic effect of reading. As Kittler writes, "poetic works, finally, are media for the hallucinatory substitution of realms of the senses"; reading became a matter of "audiovisual hallucination," of "inner possession of senses that are dead to the world."[2]

By Kittler's account, the successful transformation of poetry into a substitute for the senses brought a halt to developments in audio and visual media, and allowed literature to go unchallenged for a century as the sole medium for storing sequential sounds and images. But if one considers literature in relation not just to storage media, but also to sounds and images appearing in performance, one finds poetry not suppressing other audiovisual media but rather inspiring experiments in the production and combination of music and pictures. Celeste Langan has found as much in the case of *The Lay of the Last Minstrel* (1805) by Walter Scott, for whom Bürger's ballads (especially "Lenore") were a formative influence.[3] As Langan suggests, the *Lay* both made poetry into a "magical repository of audiovisual experience" and gave rise to numerous multimedia forms, including illustrated editions and melodramatic settings that interpreted the poem's address to eye and ear.

The Regensburg shadow-play likewise exemplifies a productive interaction between poetry, music, and moving images. It also offers a lesson in musical listening by directing listener engagement away from the elements of composition and performance to the audiovisual experience they might conjure for the imagination. Bürger aimed for his ballad to be highly sensory, and readers' accounts attest to his success. Early hearers of André's setting, by contrast, wrote not of audiovisual experiences but rather of compositional, musical, or interpretive skill. Tracing the transformations of Bürger's ballad from text to song to shadow-play thus illuminates the intertwined cultural processes by which the estimation of strophic melody changed from transparent medium of poetry to "cold and dry," while through-composed settings transitioned from distractingly artful into "visible pictures."[4]

Shadow-Plays

The idea of shadow performance is ancient, with highly developed forms being established in China and Southeast Asia at least 2,000 years ago. In Western Europe, there is evidence for shadow performances taking place as early as the mid-seventeenth century.[5] These performances, however, were largely one-off events, with no sense of shared aesthetic or technique—no sense of forming a tradition or genre. They could be stand-alone shows or take place within other presentations; they might use marionettes, paper silhouettes, or human actors performing behind the screen; and the shadows might be voiced or their mute actions narrated.

In the eighteenth century, then, shadows were at once very old and available to form a new kind of theater. This dual status is illustrated by the comments of Baron Friedrich Melchior von Grimm, a littérateur born in Regensburg and educated in Leipzig before he joined the intellectual circles of Paris in 1746. When a publication appeared in 1770 touting a "comedy for shadows, with changing scenes" performed for the first time at a private chateau in 1767, Grimm mocked the announcement. Whereas the French thought they had just invented a new entertainment, he remembered "being particularly amazed in my childhood by the noble game called *schattenspiel* in German, represented by itinerant actors with great success."[6] Both Grimm and the publication described a kind of shadow-play involving human performers moving behind the screen. For Grimm, who had published a satirical takedown of French opera as a mere mix of dances, sung speech, and vulgar chansons rather than an art of passions and grand tableaux, the shadow-play represented a similarly paltry amusement: "after the Opéra français I know of no more interesting show for children; it lends itself to enchantments, to the marvelous and to the most terrible catastrophes."[7] While these traits—enchantments, the marvelous—would supply reasons to prize shadows in a romantic context, for Grimm they represented childish attractions to be renounced in favor of true tragedy and comedy.[8]

Soon after 1770, a stable shadow-play aesthetic and technology coalesced under a new name: *ombres chinoises*. Emerging in France, this particular form of shadow-play employed silhouette puppets rather than human actors or marionettes behind the screen. While some accounts credit missionaries returning from China with importing the new form, the design and subject matter of *ombres chinoises* had little to do with Chinese models, with their colorful or intricately carved puppets and religious and historical themes.[9] More than to any Eastern origin, the "Chinese" appellation likely owed to the contemporary vogue for Chinoiserie, and capitalized on fascination with the illustrious empire and its exotic style.[10] By distancing the entertainment from previous European forms of shadow-plays, moreover, the name could have helped head off dismissive reactions like Grimm's: these were not the shadow shows of your childhood, it promised, but something altogether different from any you have encountered before.

The acknowledged inventor of the *ombres chinoises*, François Dominique Séraphin, opened his show in Versailles in 1772.[11] By the time he moved his show to Paris in 1784, it had already been copied by enterprising showmen and spread to other parts of Europe. In London, the *ombres chinoises* made their debut in 1776 under the auspices of Ambroise, an Italian (born Ambrogio) who had likely encountered Séraphin's show in France the previous year.[12] The same showman, now going by the name Ambrosio Sanquirico, brought the *ombres chinoises* to Germany in 1779, where he advertised his "never before seen here LES OMBRES CHINOISES."[13] By the 1780s, other traveling showmen too performed "ombres chinoises" throughout Germany.

The Chinese shadow-play programs of this time consisted of many short vignettes, most of a comic nature. The subjects of these vignettes came from popular stories or songs, or else centered on a particularly striking figure such as a dancer or metamorphosing magician. Especially popular were the scenarios known as "The Duck-Hunt" ("La chasse aux canards," "Die Entenjagd") and "The Broken-Bridge" ("Le pont cassé," "Die zerbrochene Brücke"), which became staples of the Chinese shadow-play repertoire and appeared on many different showmen's programs throughout Europe.

An engraving after Wilhelm Chodowiecki captures a scene from "The Broken-Bridge," and provides a sense of what Chinese shadow-plays were like in Germany around the 1780s (figure 3.1). In this comic piece, a gentle-

FIGURE 3.1. Etching by Carl Christian Glassbach after a Berlin scene by Wilhelm Chodowiecki, one of a series published in Berlin, 1790. Popular Entertainment Prints, Harvard Theatre Collection, Houghton Library, Harvard University.

man seeks information from a worker repairing the bridge. The first of his questions appears beneath the illustration: "Hey, friend! What time is it?" ("He! L'ami! . . . quelle heure est-il?"). To each question spoken by the gentleman, the worker replies evasively in song. Thus, for example, to the query "Can one cross the river?" the worker replies, "The ducks swim over it, tra-la, tra-la, tra-la."[14] The tune for these answers appears at the bottom of the engraving (the same song is recorded in an English book of airs published in 1780 by Philip Astley, where it is likewise identified with "the scene of the broken-bridge in the Ombres chinoises").[15] The resemblance between the melody's opening gesture and the later anthem "La Marseillaise" (composed in 1792) attests to its popular style. The scenario ends with the increasingly annoyed gentleman finding a boat, rowing across the river, and beating the worker for his insolence.

Published in Berlin, Chodowiecki's illustration shows a seated audience of well-groomed men and women. The orchestra, too, signals the quality of the presentation: this is not the itinerant actors' show fit for children recalled by Grimm, but a touring production for ladies and gentleman. The lights on the musicians' stands, covered by reflectors, hint at what is otherwise not obvious from the illustration: the room is darkened so that the illuminated picture may be visible. That the performance takes place in French, which was also common in England when the *ombres chinoises* were first introduced, provides another mark of distinction. As Max von Boehn has observed, the medium of the shadow-play did not gain respect in Germany until it had been imported from the cultural capital of Europe.[16]

Far from being a fait accompli, however, securing respect for the shadow theater required concerted effort. In introducing the *ombres chinoises*, Séraphin took care to differentiate his presentations from puppet shows and related forms associated with either a licentious past or mere child's play. His advertisements assured that "this show, where gaiety reigns, is still characterized by decency," such that even clerics could attend without shame; and they asserted that the shows were for everyone, not just children.[17] Wherever showmen took the *ombres chinoises*, they navigated a path between low comedy and respectability, between tired tricks and aesthetic-technological innovation.

One way to bolster the status of the shadow-play was to link it to scientific knowledge. Advertising his *ombres chinoises* in Germany, for example, Joseph Marquis identified himself not as a "director," as did Ambroise, but rather as a "mechanicus famous throughout Europe" who had brought his "famous Kunstkabinet" to the city—terms allied to practical science rather than to theater.[18] The scientific value of the shadow-play was articulated as early as the 1670s, when Gottfried Leibniz—inspired by the recent exhibition in Paris of a "machine for walking on water"—composed a short manuscript entitled "An Odd Thought Concerning a New Sort of Exhibition (or rather, an Academy of Sciences; September, 1675)."[19] "Suppose," Leibniz wrote, "that some persons of means with an interest in curiosities, especially in machines, should agree to have public expositions made of such things."[20] Included on his long list of natural wonders and artificial devices to be exhibited was a particularly detailed description of the shadow-play, which would involve

> little wooden figures, so agitated that they will throw their shadow against the paper in very startling and magnified proportions. But in order to prevent this shadow world from appearing all on one plane, resort to perspectives might bring about diminishing sizes of shadows. . . . Then of a sudden, all would be darkened. The same trick could be

> produced by subduing all lights except that one alone which is near the movable little wooden figures. This remaining light with the aid of a Magic Lantern would throw against the wall admirably beautiful and moving figures which obey the same laws of perspective. All of this would be accompanied by song from below the stage. The little figures would be moved from below or by their weight, so that whatever is used to move them is invisible. Song and music would accompany everything.[21]

Employing wooden figures rather than profiled cardboard cutouts, Leibniz's shadow-play differed technically from the later *ombres chinoises*, yet it suggests how such manipulation of light and shadow—accompanied by music—could provide more than simple diversion. Indeed, Leibniz argued that this sort of exhibition "would open people's eyes, stimulate inventions, present beautiful sights, instruct people with an endless number of useful or ingenious novelties. . . . Everybody would be aroused and, so to speak, awakened; and the enterprise might have consequences as fine and as important as could be imagined, which would some day perhaps be admired by posterity."[22] Beautiful, instructional, and inspirational, technical feats like shadow-plays promised great benefits for individual spectators, for society, and for the exhibitor who would thus come to be celebrated by future generations.

The reasoning behind Leibniz's "new sort of exhibition" carried over into the popular science of the eighteenth century. Starting with Edmé-Gilles Guyot's *Nouvelles récréations physiques et mathématiques* (1774), shadow-plays appeared in the treatises that explained nature's workings by means of hands-on experiments and amusing activities. Guyot's treatise (likely via its German translation by John Christoph Thenn) provided the source for Johann Christian Wiegleb's account of the shadow-play in his *Die natürliche Magie aus allerhand belustigenden und nützlichen Kunststücken bestehend* (1779).[23] Wiegleb made only one modification to the earlier French explanation—but a significant one, for it marks the establishment of the *ombres chinoises* as a distinctive performance medium. Guyot titled the entertainment simply "shadows" (*les ombres*), remarking in a footnote that the techniques described had been used recently in a much-admired exhibition in Paris called *Ombres Chinoises*. Wiegleb omitted mention of the Parisian performance but adopted its more specific name for the entertainment, titling his article "The Chinese shadow-play" (Das chinesische Schattenspiel).[24]

The procedures these treatises describe are consistent with what can

FIGURE 3.2. "Das Chinesische Schattenspiel" from Johann Georg Krünitz, *Oeconomische Encyclopädie, oder, Allgemeines System der Staats- Stadt- Haus- und Landwirthschaft*, vol. 140 (Berlin: Joachim Pauli, 1825). The Bancroft Library, University of California-Berkeley, AE27.K7 v. 140.

be seen in Chodowiecki's illustration. It was likely the instructions provided in natural-magic treatises that enabled amateur productions like that at Regensburg. According to these treatises, the shadow-play required a screen of fine gauze stretched across an opening in a partition. The recommended size of this opening was about four feet long by two feet high, to be positioned at least five feet off the ground. Additional screens were then set in frames inside the opening, and painted with the landscapes or interiors in which the scenes of the shadow-play were to take place. A light strengthened by a reflective mirror, centered three to four feet behind the screen, illuminated the scenes.

The shadow figures that populated these scenes were made of pieces of cardboard or like material, hinged together in order to allow for lifelike movement (figure 3.2). Thin wires were attached to the moving parts of the figures, which terminated in rings worn around the fingers of the operator's right hand. By moving his fingers up and down, the operator could make the shadow figure gesticulate. With his left hand, the operator supported the figure on a thicker wire, with which he moved the figure forward and backward in the scene. The system was arranged so that the operators and wires remained concealed from the spectators—a point emphasized, for it was key to the desired illusion of figures moving naturally and autonomously.

Finally, the shadow figures required auditory accompaniment. As Guyot and his copiers explained:

> It is essential when making [the figures] move to do some dialogue that follows their gestures and movements exactly, and one even has to imitate noises (when appropriate), that is, if one drops a figure down a ladder, one must imitate the sound that the fall of a ladder makes.[25]

The *ombres chinoises* thus involved an aesthetic of sound-image synchronization, calculated to produce a compelling audiovisual illusion. In contrast to the muted analogues to tele-visual effects discussed in chapter 1, there was no interest here in imbuing sound with a "shadowy" quality analogous to the image. Instead, while spectators saw only shadows, they heard full-bodied voices, noises, and music.

The Regensburg performance of Bürger-André's "Lenore" took place within this emerging culture of the shadow-play. It was an amateur production, but an elaborate one for which an eight-page pamphlet was printed, the text of which was reproduced in the Berlin *Litteratur- und Theater-Zeitung*.[26] Indeed, it was in order to spread knowledge of what they considered a novel event that the anonymous author(s) of this text described the logistics of the performance, provided a scene-by-scene description of the shadow-play as it coordinated with the ballad, and included an original poem that had been recited as a postlude to the performance. And while the claim to novelty was entirely typical of Chinese shadow-plays of the time, what this shadow-play presented was significantly different and new: not a series of comic vignettes, but a serious through-composed ballad.

Even as the pamphlet stressed that nothing like the shadow-play had been presented before, however, it also figured the performance as part of a natural progression. As it explained, "Mr. Bürger's dear child and Mr. André's step-daughter, Leonore, received the latest upbringing at Regensburg, where she has been given in Chinese Shadowplay by Miss. von B . . ."[27] In this construction, "Lenore" matured as it passed from Bürger's poem, to André's song, to Miss. Von B's shadow-play. Through spatial layout and font size, the title page of the pamphlet reiterated this progression from text to song to shadow-play, with the emphasis placed on Bürger's "Lenore" as the point of origin (figure 3.3). In order to understand

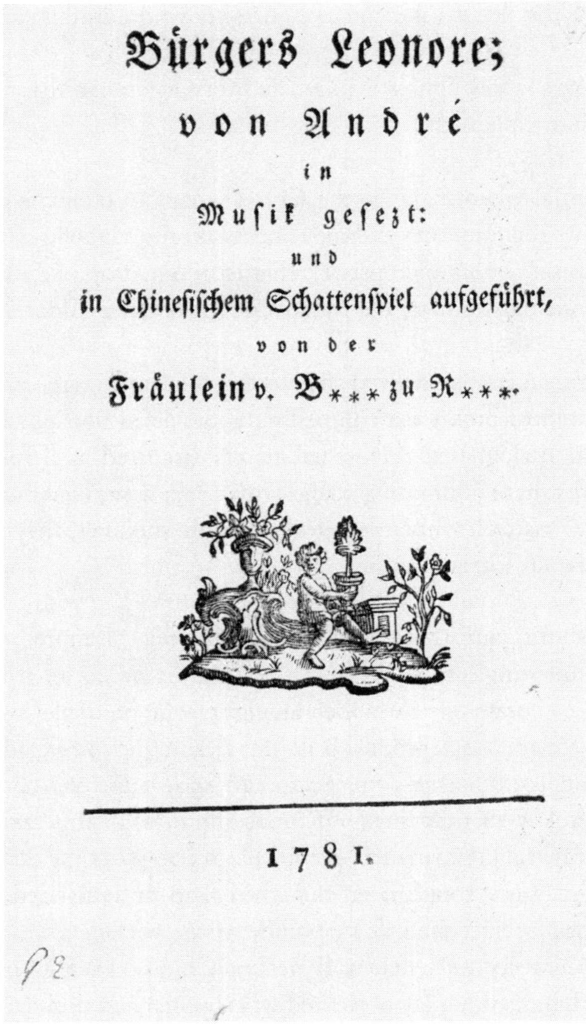

Bürgers Leonore;

von André

in

Musik gesezt:

und

in Chinesischem Schattenspiel aufgeführt,

von der

Fräulein v. B*** zu R***.

1781.

FIGURE 3.3. "Bürgers Leonore" (Regensburg, 1781). Staatsbibliothek zu Berlin, YM 1831.

the approach taken by the Regensburg performance, then, we can retrace its lineage from words, to words and music, to words, music, and moving image. This trajectory resembles traditional approaches to lied criticism in which performance stands as a means to project a composer's musical interpretation of a poem.[28] But as we shall see, the medium of the shadowplay granted performance a different status, for even as it worked to draw the experience of listening to a through-composed setting closer to that of reading the poem, it also called for an openness to the creation through performance of new meanings.

Bürger's Text

Bürger's "Lenore" tells the tale of a woman's harrowing nighttime horseback ride with her dead lover. Unfolding in thirty-two eight-line stanzas, the story of "Lenore" takes place at the end of the Seven Years War (which is to say, in recent times: the war concluded only ten years before Bürger penned the poem). The returning soldiers are welcomed home with celebratory fanfare—but Lenore's lover Wilhelm is not among them. No one can say what has happened to him, and Lenore falls into despair. Though her mother counsels her to trust in God, Lenore renounces the deity deaf to her prayers, and wishes for death to release her from life's now painful light. That night, Wilhelm unexpectedly returns. Though the hour is late and Lenore bids him stay, he insists on rushing a hundred miles by horseback to what he promises will be their wedding bed. It is a great distance to travel before the morning, but as Wilhelm assures Lenore, "The moon shines bright/We and the dead ride swiftly" (Der Mond scheint hell/Wir und die Toten reiten schnell).[29] Oblivious to the ominous signs, Lenore joins Wilhelm on his horse. As they ride, the countryside flies past them and frightful sights appear, including a singing funeral procession and demons dancing atop a gibbet. Wilhelm's earlier assurance becomes a chilling refrain: "The moon shines bright/Hurrah! The dead ride swiftly/Do you fear, dear, the dead?" (Der Mond scheint hell!/Hurra! Die Toten reiten schnell! Graut Liebchen auch vor Toten?). Though at first Lenore answers without concern, her responses to the repeated question grow increasingly alarmed. Finally, when the couple arrives at their destination, it is not a nuptial bed but a graveyard. Wilhelm decomposes into the figure of death, the horse disappears beneath them, and Lenore falls to her grave. Dancing in circles, spirits howl a moral lesson about placing faith in God rather than renouncing Him in despair.

Published in the *Göttinger Musenalmanach* in 1774, Bürger's "Lenore" quickly became widely known, rocketing Bürger to fame and earning him the sobriquet that would appear on many a subsequent publication, "author of 'Lenore.'" The ballad partakes of the emotional intensity of the *Sturm und Drang* and the ghastly horror of the Gothic. Above all, however, its aesthetic and subject matter were rooted in the interaction of literary and folk tradition. While writing the poem, Bürger sought to perfect its popular ballad style, and claimed to have taken the subject from an old spinning-song.[30]

Bürger's work was part of a movement to reform German literature by turning poets away from foreign and ancient models, toward the world of their own place and time. Bürger's position reflects especially the influence of Johann Gottfried Herder, whose essay "Briefwechsel über Ossian und die Lieder alter Völker" [Correspondence on Ossian and the songs of ancient peoples] was foundational to the movement. Reading the essay a year before completing "Lenore," Bürger found it inspiring. "The [sound] that Herder has awoken, that has long rung in my soul, too," he wrote to *Göttinger Musenalmanach* editor Heinrich Christian Boie, "has now at last been wholly fulfilled . . . I think Lenore shall largely correspond to Herder's teachings."[31] Herder's essay, informed by an engagement with British collections of folk poetry, imagined a virtuous, uncorrupted folk culture that expressed its particular modes of perception and being through its oral arts. In the essay, Herder advocated collecting the oral poetry of the *Volk* and infusing written poetry with its powerful subjects and unbridled style. Contrasting the power, immediacy, vivacity, and naturalness of folk songs with the restraint, coldness, lifelessness, and artificiality of poetry written for the page, Herder emphasized the "living presentness of the imagery" in folk songs—a visuality that proceeded from the folk mentality, uncontaminated by rationalism or abstraction. As Herder illustrated the point, "when the Greenlander speaks of his seal hunt, he does not narrate, but he paints; he paints each circumstance and each movement with both words and movements, for they are all part of the image that exists in his soul."[32] It was also this vivid quality, Herder argued, that made a sensuous imprint on the memory and enabled oral poetry to survive without recourse to dead paper and ink (nor even—as David Wellbery points out—to the mnemotechnics and other performance trainings in fact practiced by oral cultures).[33]

Bürger was particularly invested in the "living presentness of the imagery" that Herder celebrated in folk poetry. William Little has observed that Bürger's correspondence reflects an "insistent concern for effect ('Wirkung'). The immediate sensory effect of a work was of towering importance to him, and again and again he stresses the impact he hopes to make."[34] Similarly, Martha Woodmansee has found that Bürger's "goal throughout was to bring his material alive for his readers—to enable them to imagine it as vividly as if they were actually experiencing it."[35] In drier aesthetic terms, Bürger sought a quality of vivacity that aimed literature less toward the understanding than toward the imagination. This goal was not limited to poetry meant to imitate oral folk traditions. In his general, comparative theory of painting and poetry (*Laocoon*, 1766), Gotthold Ephraim Lessing wrote that while the prose writer may be satisfied merely with being intelligible, the poet "wants to make the ideas he excites in us so vivid, that from the rapidity with which they arise we believe we are feeling the genuine sensate impressions of the objects of those ideas; and in this moment of illusion we cease to be conscious of the means he employs to this end, namely his words."[36] Johann-Christoph Adelung, in a treatise on German writing style, described "proper vivacity" as bringing about "a condition of the force of representation in which the sensory representation predominates over the imagination and moves us more strongly than what we are sensing."[37] In other words, deploying language to stimulate illusory sensory experiences was a common poetic aim. But for Bürger, such vivacity was part of a larger social vision. As he explained in his essay, *Herzens-Ausguss über Volks-Poesie* [Confessions on popular poetry, 1776], appealing not only to the understanding but also to the imagination and senses was essential in order for poetry to reach the learned and leisured of the city as well as the untutored of the forest or farm. And it was popularity across these strata of society that was the true measure of a poem's greatness.[38]

The experiences reported by readers and hearers of "Lenore" attest to Bürger's success at achieving sensory effects. One reader wrote to his brother that the ballad "shattered my whole nervous system for an entire night, and when Bonstetten was reading it about midnight and the door suddenly sprang open, he dropped the book from his hand, and his hair stood on end."[39] For Herder, the effect was a lingering hallucinatory effect: "when I read it," he wrote, "it took hold of me so, that that afternoon I saw naked skulls on all the church pews."[40] In an essay on Bürger published in 1800, August Wilhelm Schlegel recounted Bürger's reciting the ballad aloud to his friend Count Friedrich Leopold Stollberg, soon after having completed it. When he reached the lines

> Rapidly they came
> Full gallop to an iron gate
> With swinging whip a blow to it
> Burst lock and key,

Bürger snapped a riding whip against a door on the opposite side of the room, causing Stollberg to "[jump] with horror, as if the things described were really going on before his eyes."[41] Schlegel identified literary techniques that made the ballad so vivid, describing "Lenore" as unfolding its story of emotional strife and horrible death "in a few easily comprehensible lines and living, flying-past pictures . . . and almost entirely in alternating speeches, during which one sees the figures move and gesture without the assistance of disturbing descriptions."[42] To the simplicity of language and predominance of dialogues over description, literary scholars have added the rhythms and onomatopoeic sound effects of Bürger's language, which according to Woodmansee cooperate with the visual imagery "to produce a tension difficult to surpass before the invention of moving pictures."[43]

By Kittler's chronology, the period of "Lenore's" popularity spanned the shift from old to new reading pedagogies—from the production of audio-visual experiences only as a special poetic effect, to their reliable occurrence as an automatic result of reading. In this light, the prevalence of stories about being shocked by an event in the environment registers not only the fearfulness of the poem but also the newness of its grip on the imagination: these events tested readers' relative awareness of their surrounding world and the things represented in the poem, and revealed the extent to which they had become oblivious to the former and immersed in the latter. The case of Bürger's "Lenore" also suggests that Kittler underestimated the contribution of poetic works to changes in reading practices. Whether recited aloud or read silently, "Lenore" fostered a new type of literary experience focused on sensory illusion.

If vivid imagery was thought to enable poems to persist in memory, equally important was their setting in song. Though early collectors such as Herder typically recorded only texts, these texts were imagined to exist off the page as songs. For the literary imitators of folk poetry, having their words united with melody was a common dream, connected to the hope that their poems would reach beyond bourgeois readers to become part of an oral folk culture. As Bürger wrote to Boie while at work on "Lenore": "My greatest reward would be to have it set to music simply and in a true ballad-manner, and again used at the spinning wheel."[44] The "again," here, is significant. Bürger not only claimed to have taken the subject of "Lenore" from an old spinning-song but also dreamed of its return—from its new literary circles—to its folk source.

Song as Medium

At the time when Bürger expressed the hope that his poem would be set to music and again sung at the spinning wheel, composers too were developing a vision for simple songs accessible to all. Already in the first half of the century, opera had become a battleground for foreign (Italian) versus native (German) elements—the former inevitably coded as "artificial," the latter as "natural." In the 1770s, as we saw in the previous chapter, librettists and composers developed a new brand of opera for the German people; they also began to formulate native elements explicitly in terms of the *Volk*. Abandoning the virtuosic Italian style of the da capo aria for simple melody and strophic forms, Johann Adam Hiller and librettist Christian Felix Weisse made comic opera a platform for the dissemination of songs in folk-tone. As Weisse, in 1778, wrote of their collaborations, "no medium can . . . be more powerful in making song more universal than comic opera . . . All the songs which were well liked in performance were soon sung by all, at first bringing enjoyment to social gatherings, and then even going over to the ordinary *Volk*."[45]

This vision—of songs composed in the folk style moving from educated, bourgeois circles to "the ordinary *Volk*"—typified the social-aesthetic program of many lied composers in the 1780s. These composers published collections of songs with simple or even optional keyboard accompaniment, and emphasized the qualities of accessibility and memorability they sought for their melodies. Johann Abraham Peter Schulz, for example, explained in the preface to his second volume of *Lieder im Volkston* (1785) that he aimed to write melodies so simple and natural that even the untutored could sing and remember them. To that end, he strove "for the greatest simplicity and comprehensibility" in his melodies, seeking to attain "the appearance of the unforced, the artless, the familiar, in a word, the *Volkston*, whereby [the lied] impresses itself on the ear so quickly, returning unceasingly."[46] Johann Friedrich Reichardt, aiming for songs accessible to anyone with "ears and a throat," explained that the melody must be so simple and matched to the text

> that once one knows the melody, one will no longer be able to think of it without the words, nor the words without the melody; that the melody will be everything for the words and nothing for itself.
>
> Such a melody . . . will always have the character of a unison, and thus require no accompanying harmony or else allow it only as a concession.[47]

This self-effacing musical aesthetic was nonetheless paired with claims for creative excellence and freedom for the composer, and a commitment to leading German authors. Composers applied their skills to the "best" poetry, selecting texts by celebrated figures such as Voss, Klopstock, Goethe and, of course, Bürger.[48]

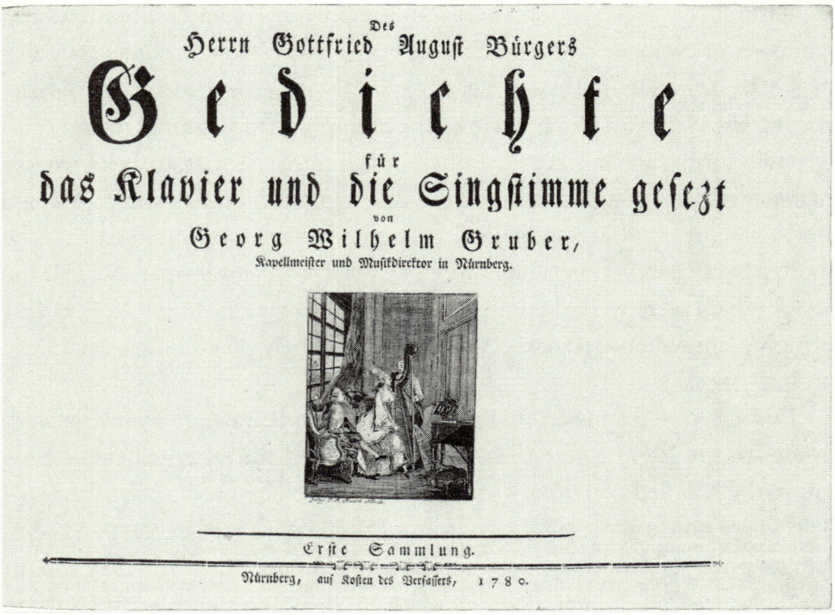
Des
Herrn Gottfried August Bürgers
Gedichte
für
das Klavier und die Singstimme gesezt
von
Georg Wilhelm Gruber,
Kapellmeister und Musikdirektor in Nürnberg.

Erste Sammlung.
Nürnberg, auf Kosten des Verfassers, 1780.

FIGURE 3.4. Georg Wilhelm Gruber, *Bürger's Gedichte für das Klavier und die Singstimme gesezt* (Nuremberg, 1780), title page. SUB Goettingen, 8 MUS VII, 790:1.

By 1781, Bürger's "Lenore" had appeared in no fewer than four collections of songs, receiving settings by Friedrich Wilhelm Weis, Georg Wilhelm Gruber, Johann Philip Kirnberger, and Georg Friedrich Wolf.[49] While differing in the manner or degree to which they pursued the folk-song aesthetic, these settings share a strophic approach to the text, and a view of literate, domestic performance as their first, if not only, destination. Gruber's setting, which appeared in a song collection devoted entirely to Bürger's poetry, is the least calibrated for dissemination beyond literate circles. A Kapellmeister in Nuremberg, Gruber worked at a remove from the Northern German centers in which the aesthetic-social program of folk song primarily developed. An illustration on the title page shows the envisioned performance setting for the publication: in a bourgeois salon, two gentlemen give their attention to a lady sitting at a harp, a keyboard standing ready at their side (figure 3.4). The ornamentation of the vocal line (both written out, as with the slurred sixteenth note in mm. 12–13, and indicated by symbols, like the turn in m. 12) provides the musical refinement suited to this intended audience, as do the introductory and closing sections for keyboard alone. The triplet-sixteenth-note passages in these instrumental sections, in an otherwise duple-meter context, suggest the galloping horse.

PLATE 1. Martin Engelbrecht (1684–1756), "Presentation of an Opera with Hovering Figures," paper theater (*Kulissentheater*), Augsburg, eighteenth century. Above: lower right, arrangement of cards as would appear inside a peepbox; lower left, view with resulting depth effect. Below: close-up view of cards stacked atop one another. Theatermuseum, Vienna.

PLATE 2. Itinerant showman's peepbox, central Europe, second half eighteenth century. Erkki Huhtamo Collection, Los Angeles. Photo: David Leonard.

PLATE 3. E. T. A. Hoffmann, illustration for Friedrich Baron de la Motte Fouqué, "Der Kuckkasten" [The peepbox], in [K. W. S.] Contessa, Friedrich Baron de la Motte Fouqué and E. T. A. Hoffmann, *Kinder-Maehrchen* (Berlin, 1817). Houghton Library, Harvard University, GC8.C7673.816k.

PLATE 4. John Scott, magic-lantern slide, London, late eighteenth century. Cinémathèque française.

The key of D minor and descending chromatic lines prominent throughout the song, meanwhile, paint the dark scene and ominous mood. In the preface to the publication, Gruber explained that it was at the request of some subscribers that he added keyboard introductions and conclusions to most of the songs in the collection.[50] Evidently, his literate audience had a taste for more elaborate musical settings of Bürger's poems—but tastes still accommodated by strophic forms (example 3.1).

Kirnberger's setting represents an opposite extreme of simplicity and apparent uninterest in musical expression. Kirnberger published his setting of "Lenore" while in the service of Princess Anna Amalia in Berlin, in a collection titled *Gesänge am Clavier* (1780). Kirnberger's choice of major strikes an immediate difference from Gruber's ominous minor; his melody is also more singable, the range being confined to an octave and the motion being entirely stepwise or by thirds or fourths. The accompaniment makes no hint at illustrating the text. The repeated sixteenth notes in the bass and almost constant eighth notes in the melody give the song a uniform texture. Moreover, the same music is repeated for lines 1–2 and 3–4. The song is thus highly consistent—a critic might say monotonous—and the even, stepwise melody has a carefree air that finds no correspondence in the text (example 3.2).

Though Kirnberger's setting seems most foreign to modern sensibilities, it is likely closest to what Bürger had in mind when he hoped his text would be set in "a true ballad-manner." For it was with such simple melodies, reusable for various texts, that ballad singers regaled the public with unsettling stories at fairs and other outdoor venues. A means of disseminating news as much as a form of entertainment, ballad singing dates to the sixteenth century. In the eighteenth century, with the rise of other media for public discourse, educated people sought to distance themselves from the lowly ballad tradition. Critics often lumped ballad singers together with itinerants of dubious means and little merit, such as rope-dancers, beggars, and gypsies. The miraculous tales ballad singers told also made them the target of criticism. Georg Wilhelm Wegner's anti-superstition treatise *Schauplatz Vieler Ungereimten Meynungen und Erzehlungen* (1735), for instance, complained that false beliefs were promulgated by "the vagabonds who travel from market to market and recite all sorts of sad murder-stories, the better to sell their printed songs telling these stories; and who even go further and have with them paintings of such shocking stories, and while reciting, point at them with a stick, in order that the simple people may be more inclined to believe them."[51]

EXAMPLE 3.1. Georg Wilhelm Gruber, "Lenore," *Bürger's Gedichte für das Klavier und die Singstimme gesezt* (Nuremberg, 1780)

Lenore stirred at first dawn red/From out her dreams so deep:/"Art untrue, Wilhelm, or art dead?/How long your tarry keep?"/Abroad with King Frederick's might/He'd entered into the Prague fight,/And he had never written/To tell if he remained healthy.

EXAMPLE 3.2. Johann Kirnberger, "Lenore," *Gesänge am Clavier* (Berlin und Leipzig: George Jacob Decker, 1780)

As this complaint attests, ballad singers did not just sing. They made their meager earnings mainly from the sale of printed texts—cheap broadsides or chapbooks that the performer commissioned, or else a printer produced who then recruited the singer.[52] In addition to singing the texts of these prints, the ballad singer also illustrated them on large posters. These posters typically displayed small pictures arranged on a grid, with each picture corresponding to a strophe of the ballad (figure 3.5). The poster served both to advertise the ballad singer, drawing spectators to him, and as an object of interest in its own right.

FIGURE 3.5. Christian Wilhelm Ernst Dietrich, "Der Bänkelsänger" [The ballad singer] (print in the style of Adriaen van Ostade), 1740. The Getty Research Institute, Los Angeles (2011.PR.34).

The resulting ballad picture show was a form of public performance that mixed speech, song, gesture, and painting to relate dramatic tales of crime and punishment, miraculous events, or catastrophes. As Tom Cheesman has shown, ballad singers typically began a performance by narrating a story in prose, then sang the same story in verse while pointing to the corresponding pictures. According to Cheesman, the didactic structure of the ballad picture show, with its repetition, illustration, and pointing, made it a means of not only teaching particular stories and moral lessons but "probably *reading itself*."[53] Ballad singers sold prints even to illiterate spectators, who might correlate their texts with their memory of the oral performances.

In the 1770s, the new interest in folk song reawakened interest in ballad singers.[54] But they remained lowly figures, their commercially motivated hybrid of oral and literate forms being incompatible with the idealized vision of the folk. The ambivalent attitude toward these vagabonds also found expression in the vexed status of visual attractions—for while a strong visuality might signify the primal virtues of the folk, it might also signify an inferior intellect. Reflecting on the folk sensorium, for example, Goethe wrote disapprovingly that "the Volk is stirred the strongest by all that is brought under the eyes. Far more than a detailed description, a botched picture or childish woodcut draws the dark man . . . The large pictures of the ballad-singers make a far deeper impression than their songs, though these too grip the imagination in strong fetters."[55]

By what means song gripped the imagination—through what relationship of text and music—became a preoccupying question as poets and composers moved beyond the issues of accessibility and memorability. Goethe, who often discussed the matter with the composer Carl Friedrich Zelter over their thirty-five-year correspondence, considered strophic settings with subtle, occasional word-painting best at enhancing the experience of the text. As Goethe wrote (assigning more credit to the text for stimulating listeners' vision than did Hiller, whose theory of music inducing vision was discussed in the previous chapter), "the object is to transport the listener into the frame of mind which the poem itself suggests; the imagination will then form pictures from the text without knowing how it arrives at them."[56] Competing approaches to song composition fueled such critical focus on music's role in stimulating the imagination's picture-forming capacities. For at the same time that poets and composers articulated the values of simple folk-style songs, and various strophic song settings multiplied, more elaborate approaches were also percolating.

Through-Composition

Johann André published his first version of *Lenore* in 1775, not in a collection of songs but as a stand-alone composition (a performance of the piece can last nearly half an hour). Reviewing this publication in his *Deutsche*

Chronik, musician and poet Christian Friedrich Daniel Schubart highlighted the significance of André's approach to the ballad:

> [Bürger's] Lenore is such an excellent piece of poetry, so full of tone and musicality, that it was a wonder to me why no skillful master had yet set it to music. [Ignaz von] Beeke, that great practical and theoretical musician, set it in the true ballad-tone. But since all the strophes have a single melody, he could not exhaust all the beauties of the original poem. Mr. André is thus the first to set the whole of Lenore verse by verse in a way that we easily see that he could empathize with the poet; only he does not appear to have enough musical art, as one finds here and there very objectionable places.[57]

André may not have been the consummate master for which Schubart hoped, yet he performed the critical role of occasioning direct comparison between strophic and through-composed approaches to the ballad—prompting a favorable light to be cast on the latter.

André was both a publisher and a composer, though his musical training was largely sporadic and informal. Prior to setting "Lenore," his first opera, *Der Töpfer* [The potter, 1773]—a comic one-act to his own libretto—won him public approval along with comparisons to Hiller for his simple melodic style. Having established a publishing business at Offenbach in 1774, he also tried his hand at lied composition, producing two self-published collections in his firm's first year: *Scherzhafte Lieder* [Humorous songs] and *Auserlesene scherzhafte und zärtliche Lieder* [Exquisite humorous and tender songs]. The former set lighthearted texts from Weisse's poetry collection of the same name. With the latter, André added to the standard voice and keyboard parts accompaniments for flute, violin, and bass.

In a review of the *Auserlesene scherzhafte und zärtliche Lieder*, Schubart praised both the easy and pleasant quality of André's melodies, which "impress themselves on the memory," and their multi-instrumental arrangements, which were suited for "agreeable family concerts on Winter evenings" where "the daughter plays the keyboard and sings; the brother takes the flute or violin, and the father strikes the bass."[58] In these two assessments, one can see Schubart balancing folk and bourgeois ideals. According to Schubart, and consistent with Herder's ideas, true folk songs were unsophisticated, easy, and "flowed freely from the emotional heart via paths dug by nature—not via canals picked and shoveled by [Berlin composer-theorists like] Marpurg and Kirnberger."[59] Schubart criticized Berlin lied composers, whose songs he claimed were scarcely sung or liked anywhere due to their being too plodding, and written with an anxious conscientiousness.[60] But far from insisting on an artless simplicity, Schubart—who studied music with Gruber in Nuremberg in the 1750s—valued embellishments that increased the music's interest for literate amateurs.

Shortly after his first song collections—and within months of Bürger's poem appearing in the *Göttinger Musenalmanach*—André self-published his setting of "Lenore." While Schubart had been concerned primarily with folk ethos in his review of André's *Auserlesene scherzhafte und zärtliche Lieder*, with *Lenore* he directed his critical focus to adherence to the rules of music, and accurate expression of the poet's ideas. If André's compositional skill was suspect in his own time, his reputation can suffer even more by comparison with later composers of lied. Marjorie Hirsch, for example, has written that the keyboard part of André's first *Lenore* publication "does little more than provide harmonic support for the voice."[61] But such a retrospective stance misses not only the virtues of simple accompaniments, but also André's quite elaborate use of the keyboard to express the poem. Indeed, the differentiation of musical sections (by tempo, key, meter, phrasing, etc.) for dramatic purpose (for instance to distinguish characters or highlight certain events) is to be found not only in Franz Schubert's dramatic ballads but also in André's *Lenore*.[62]

Schubart's comment that André set "Lenore" "verse by verse" aptly captures not only his through-composition of the text, but also his sectional approach to the composition. Every stanza ends with a full cadence, often followed by a rest; almost every stanza begins with a new key and tempo; and most stanzas feature a musical character (expressed through tempo, keyboard figuration, etc.) that contrasts with those of the stanzas immediately preceding and following. Table 3.1 summarizes the tempo and key areas of the stanzas. As can be seen from the table, the changes in tempo from one stanza to the next are sometimes extreme (Un poco allegro to Grave in stanzas 20 to 21, for example). Through this approach, André preserves the sectional, stanzaic structure that in strophic settings comes from the repetition of the same music.

Because André follows the text's stanzaic structure as a rule, it is particularly significant when he departs from that structure, whether by linking stanzas through tempo and/or key relationship, or by creating subsections within a stanza. These departures typically come from following the characters' dialogue. When characters exchange dialogue within a stanza, the different speakers are often marked with distinct keys, tempos, and/or figu-

TABLE 3.1. Structure of André's *Lenore* (1st ed.)

STANZA	TEMPO	KEY	SPEAKER
1	Andante	G	Narrator
2, 3 (repeat)	Allegretto	B-flat	Narrator
4	Un poco lento	b-flat	Narrator
5	Moderato–Allegro	f	Narrator–Mother–Lenore
6, 7 (repeat)	Larghetto–Allegro	A-flat–c	Mother–Lenore
8, 10 (repeat)	Andante	E-flat–g–E-flat	Mother
9, 11 (repeat)	Allegro	c	Lenore
12	Andantino	F	Narrator
13	Un poco piu lento	B-flat–g	Narrator
14	Moderato–Allegretto	g–B-flat	Wilhelm–Lenore
–15		c–g	Wilhelm–Lenore
16	Allegro assai	g–B-flat–c	Wilhelm
–17		c–g	Lenore–Wilhelm
–18	–Adagio–Tempo di primo	G–d–g	Lenore–Wilhelm–Lenore–Wilhelm
19	Allegretto	B-flat	Narrator
20	Un poco allegro	E-flat–b-flat–E-flat	Narrator–Wilhelm–Lenore
21	Grave	C–E-flat–C	Narrator–Spirit Chorus–Narrator
22	Piu vivo	c	Wilhelm
–23	–Allegretto	–E-flat–c	Narrator
24	Allegro	C–c–C	Narrator–Wilhelm–Lenore
25–26	Allegretto	c–E-flat–c	Narrator–Wilhelm–Narrator
27	Allegro molto–Adagio	C–c	Narrator
28	Vivace–Lento–Andantino	C–c	Wilhelm
29	Presto	f–A-flat–f	Narrator
30	Moderato	b-flat	Narrator
31	Allegro assai	B-flat–f–B-flat	Narrator
32	Allegretto fugato	g	Narrator–Spirit Chorus

rations (see table 3.1). In stanza 6, for example, Lenore's mother prays for God's help at a *larghetto* tempo in Ab major and chorale style, while Lenore rejects her mother's prayers and God at *allegro* and in C minor, with repeated eighth notes in the bass helping express her agitation and impatience (example 3.3). Stanza 20 is *un poco allegro* throughout, but the narrator sings in E-flat major in constant triplets (mm. 556–563), Wilhelm

EXAMPLE 3.3. Johann André, *Lenore* (Offenbach am Main: Johann André, 1775), mm. 82–97

"Help, God, help! Look mercifully upon us!/Child, pray to the father above!/What God does, he does from love;/God have mercy on us!"/"O Mother! Mother! Vain delusion!/God has not treated me well!/What help were my prayers?/Now there is nothing left to need!"

"Help, God, help! Whoever knows the Father/knows that he helps his children./The holy Sacrament/will relieve your misery."/"O Mother! Mother! What burns me/no Sacrament can soothe. No Sacrament can give back/life to the dead!"

sings in B-flat minor over an oscillating half-step figure in the bass (mm. 564–579), and Lenore concludes the stanza in E-flat major in recitative style (example 3.4).

While changes in key, texture, and/or tempo within stanzas articulate additional internal sections, repeated alternation between musical areas can also function to connect multiple stanzas into a more continuous whole. André uses this technique to establish a dialogic back-and-forth between speakers. Such multi-stanza complexes occur in two places in André's setting (shaded in table 3.1). In the dialogue between Lenore and her mother (stanzas 5–11), Lenore consistently sings in minor over an agitated accompaniment while her mother sings chorale-like material in major. In the dialogue between Wilhelm and Lenore (stanzas 14–18), Wilhelm sings predominantly in minor mode throughout with recurring musical motifs, while Lenore goes from highly contrastive material in Bb major (her initial delight at Wilhelm's return) to minor material increasingly akin to Wilhelm's (her growing dismay and recognition of what is happening).

EXAMPLE 3.4. André, *Lenore*, mm. 556–584

To the right and to the left/before their very eyes/how flew the heaths, meadows and country!/ how the bridges thundered!/"Do you fear, dear? The moon shines bright!/Hurrah! The dead ride swiftly!/Do you fear, dear, the dead?"/"Oh no! Forget the dead!"

Schubart had special praise for André's expressive and imitative devices—particularly when they were frightening. He described the dialogue between Lenore and Wilhelm in stanzas 14–18 as "horrifying" (*schreckliche*), and among the most successful parts of the setting. Here, André employed musical signifiers for the oracular and the supernatural in Wilhelm's low octave-unison accompaniment and monotone lines (examples 3.5a and 3.5b).[63] Schubart also praised André's "frightful" (*fürchterlich*) setting of Wilhelm's refrain "Do you fear, dear?" over an oscillating half-step bass (see example 3.4 above, mm. 564–572); the funeral song "Let us bury the body," which André set to a drawn-out circular melody (example 3.6); and the spirits' words in the final stanza, which André set to an inverted sigh figure with surging dynamics, and which Schubart thought "expressed the death-howls of the spirit very well" (example 3.7).[64]

Despite Schubart's prognostication, André's *Lenore* seems rarely to have made listeners shiver. After hearing a performance of André's *Lenore* at the house of poet Heinrich Wilhelm von Gerstenberg in Lübeck, Johann Erich Biester wrote to Bürger: "Do you already know your Lenore with music by André? . . . O Bürger! Bürger! Such glory of music, such power of

EXAMPLE 3.5A. André, *Lenore*, mm. 305–320

We saddle up only at midnight./I rode from far Bohemia;/I started on my way so late/and want to take you with me.

EXAMPLE 3.5B. André, *Lenore*, mm. 354–369

Let it blow through the hawthorn/Let it blow, child, let it blow!

EXAMPLE 3.6. André, *Lenore*, mm. 599–607

Let us bury the body!

EXAMPLE 3.7. André, *Lenore*, mm. 937–957

Patience! Patience! Even when your heart would break!/No quarrel with God in Heaven make!/Thy body has played out its role;/God have mercy on your soul!

song! How each idea is completely taken and completely expressed! Truthfully! Naturally! Some places are admirable beyond all expression. How it has refreshed my heart!"[65] Instead of reacting to André's supernatural music, Biester evaluated how well the music expressed the poem, reacting with joy to its appropriateness. He also directed attention to the skill of the performers, effusing that "you won't believe what superior musical talents both Gerstenbergs have. She plays the piano with eager passion, and has a gentle, beautiful, melodious, flexible, expressive voice . . . She played, and she and he sang, for Lenore is set as a duet [*sic*]."[66] Rather than becoming immersed in the "living presentness of the imagery" and responding to the frightful occurrences, Biester remained engaged with his surrounding environment and reacted to the artistry of composer and performers. Attentive to their skill and admirable expression of the text, Biester experienced not a shattered nervous system but refreshment of the heart.

Biester's focus was typical of the kind of social space in which André's "Lenore" was usually heard. These were gatherings in which conversation coexisted with performances of poetry and music. André himself frequently performed his "Lenore" at such gatherings, as Goethe recalled in his autobiography:

> Bürger's "Lenore," then but just published and received with enthusiasm by the Germans, had been set to music by him [André]; this piece he was always forward to execute however often it might be encored.
>
> I too, who was in the habit of repeating pieces of poetry with animation, was always ready to recite it. Our friends at this time did not get weary of the constant repetition of the same thing. When the company had their choice which of us they would rather hear, the decision was often in my favor.[67]

In what measure the company favored Goethe as a performer, or preferred the verbal to the musico-verbal presentation of the poem, Goethe does not say.[68] But with poetic recitation and musical performance taking place in the same social spaces on the same texts—and with choosing between the two a part of the evening's activities—comparison was inevitable. Such direct comparison encouraged a focus on the differences between the two modes of presenting a poem—particularly, on the additional layers of artistic mediation that musical performance involved.

The distractions of musical art formed a main pillar of the defense of strophic settings against through-compositions. Goethe, for example, found that actor and singer Wilhelm Ehlers's ballad performances "[brought] out to one melody the most various meanings of the single verses," and demonstrated "how despicable is all so-called through-composition of songs, by which the general lyric character is quite effaced, and a false sympathy in details furthered and excited."[69] In other words, through-composition destroyed the unity of the whole, and substituted musical elaboration for the singer's direct expressivity. The musicians in Bürger's circle seem to have harbored similar commitments to strophic song, telling the poet that André's setting of "Lenore" was execrable.[70] But if removing layers of artistry was one way to return to the "living presentness" of the ballad's imagery, so too was adding more layers—as happened when André's "Lenore" was performed as a shadow-play.

In Chinese Shadow-Play, "For Pleasure as Well as Horror"

The "Lenore" shadow-play took part in a wider practice of playful performance, in which works of literature or painting were enacted in tableaux vivants, charades, amateur theatrical productions, and the like.[71] Such practices flourished in German salon culture from about 1780, joining poetic recitation and musical performance to enliven social gatherings where conversation ruled. In general, there is less information available about these private events and their performances than about their public, professional counterparts. If Regensburg was home to "open house" salons where "uninvited" friends were welcome, as they were at Henriette Herz's famous salon in Berlin, they are today unknown.[72] The "Lenore" shadow-play seems to have been publicized only after the fact, in the *Litteratur- und Theater-Zeitung* article that reproduced the text of a specially printed pamphlet. While this text tells us the shadow-play was "performed at night/for pleasure as well as horror" (gabs bey Nacht/Zur Lust, wie auch zum Grausen), it says nothing about who attended—and even its organizer's and performers' identities remain veiled.[73]

The pamphlet credits a certain "Fräulein v. B*** of R***" with directing the Chinese shadow-play, explaining that she not only mounted the performance of Bürger's ballad "according to her own conception and elaboration, with music,"[74] but also sang the part of Lenore. Thanks to her family archive and connection to the composer André, "Fräulein v. B***" can be identified as Regensburg resident Henriette von Berberich. When organizing the shadow-play performance, Henriette was a teenager and an eligible young bachelorette. Her father, Baron Franz Ludwig von Berberich, was privy councilor to the noble Thurn und Taxis family, which had risen to wealth and social prominence by establishing an international postal system. In 1786, five years after her shadow-play, Henriette married Alexander Conrad von Vrints-Berberich (1764–1843), who served as postmaster general in the Thurn and Taxis realms. For at least another decade, Henriette remained active in music and the visual arts. She painted miniatures, copying the work of old masters.[75] And her musical talents earned her multiple publication dedications, including one from André, who named her the dedicatee of his fourth version of "Lenore" in 1790.[76] The Fürst Thurn und Taxis Hofbibliothek und Centralarchiv now includes the former private library of Alexander and Henriette, a collection whose roughly 6,000 items—many of them German and French literary works from the late eighteenth and early nineteenth centuries—attest to the couple's literary interests.

Unfortunately, no musical source material for the shadow-play performance survives, though this point requires elaboration. The Fürst Thurn und Taxis Hofbibliothek und Centralarchiv counts among its holdings a print of the third version of André's "Lenore," scored for four solo voices, four-part chorus, and orchestra (two flutes, two oboes, two bassoons, two horns, and string quintet), plus a set of manuscript parts copied from the print. The catalog entry for these parts identifies them as "performed as a shadow-play in Regensburg in 1781" (1781 in Regensburg als Schattenspiel aufgeführt). These parts cannot in fact have been used in the 1781 performance, however. First, André did not publish this version of "Lenore" until 1788. Second, the shadow-play performance did not make use of wind instruments or a string quintet, but rather employed a Späth keyboard instrument and "outstanding English fortepiano."[77] As André's second version of "Lenore" was not published until 1782, the shadow-play was most likely performed from the first edition, though the print seems not to have been preserved in the library collection.

With its "shadows really speaking and acting," the "Lenore" shadow-play aspired to bring the ballad to life—to put its happenings directly before spectators' eyes and ears. To this end, each character in the shadow-play was voiced by a different singer. While three-inch-high shadow figures of Lenore with her mother or "the horseman" (Wilhelm) moved about onscreen, the singers and keyboard players performed in the dark, using shielded lanterns so they could see their music without interfering with the illumination of the shadow-play. In addition to the individual voice parts for the characters' dialogue, several portions of the text were sung by mixed chorus: the mother's prayers "God help us . . . ," the song of the funeral procession that Lenore and Wilhelm ride past, and the concluding moral lesson sung by the spirits over Lenore's grave.

The narrative parts of the poem were left to a fourth solo singer, who appeared not as a shadow figure but in person as a "ballad singer." Standing in front of the shadow-play in the same darkened room as the spectators and gesticulating with a white rod, the ballad singer framed the shadow-play as a ballad picture show of the type familiar from streets and fairgrounds.[78] While ballad singers were usually viewed as anonymous itinerants performing authorless songs, the narrator of the shadow-play appeared "dressed in the costume of Hans Sachs," the famous sixteenth-century *Meistersinger*.[79] Ignored for much of the eighteenth century, Hans Sachs's star had risen along with that of folk poetry in the 1770s. Goethe rediscovered Sachs in this period, and on the bicentenary of his death in 1776 published the poem "Explanation of an old woodcut depicting Hans Sachs's poetic calling" (Erklärung eines alten Holzschnittes, vorstellend Hans Sachsens poetische Sendung).[80] Though no such woodcut existed, the poem described a muse bestowing on Sachs the power to perceive the truth, and to recount all sorts of stories—biblical, historical—as if he had actually witnessed them.

By his presence before the shadow-play, the ballad singer embedded an innovative spectacle within a traditional frame. The persona of Hans Sachs, moreover, placed the scene of performance in the distant past. Perhaps resonating with Herder's teachings about the vivacity of ancient oral poetry, the costume suggested that to recover the art of the *Meistersinger* was to recover the ability to make living pictures appear before the listener's eyes. At the same time, the recourse to technological spectacle could be taken to suggest the fallen condition of modern man, who requires such artificial stimulation to compensate for his deadened imagination.

The report provides fairly detailed descriptions of the scenes that passed behind the ballad singer's rod (see table 3.2). From these, we can see other ways in which Berberich adapted André-Bürger's "Lenore" for the screen. In addition to distributing the vocal part among multiple voices in keeping with the idea of "shadows really speaking and acting," the shadow-play entailed imposing a theatrical structure upon the ballad, dividing the text into acts to accommodate scene changes; and it required resolving ambiguous, impractical, or suggestive poetic descriptions into pictorial specificities.

As table 3.2 shows, Berberich divided the ballad into five acts; at the end of each act, a curtain fell and the scenery was changed for the next act (table 3.2 puts scenery descriptions in italics). All act and scene changes lined up with the division of the text into stanzas, but these stanzas were unevenly distributed among the acts: act 4, for instance, has only two scenes across three stanzas, whereas act 1 has four scenes across eleven stanzas. This unevenness likely occasioned the report to observe defensively that "the acts and scenes of the piece could not be arranged according to the rules of art, but rather according to the decorations, as sketched out by the romance."[81] In other words, the logistical requirements of scene changes—not matters of dramatic structure, pacing, or proportion—governed the division of the ballad into acts.

The description of scenes shows attention to visual details in Bürger's poem, as well as added details and alterations made for the sake of the

TABLE 3.2. Description of scenes from the Regensburg shadow-play report

STANZA	ACT, SCENE	
1	1, 1	*The stage presents a rural area. On the street stands a single farmhouse.* The ballad singer narrates . . . Leonore steps out of the house, wrings her hands, and sings . . . She goes back into the house; the ballad singer narrates.
2–3	1, 2	The ballad singer describes the return of the army; the army marches past Leonore's house. A chorus of dancing peasant lads and lasses follows it.
4	1, 3	Leonore hastens after the army and searches in vain for her lover. She falls in despair. The ballad singer describes her condition.
5–11	1, 4	The mother comes to comfort her daughter. Her ideas are fruitless, and Leonore breaks out in the most heinous curses. "Help, God, help! Look mercifully upon us," etc., and "Help, God, help! Whoever knows the Father," etc., are sung in chorale style, in four parts, in their places.
12	2, 1	*The same farmhouse from another side, where stand hedges and shrubs. It is night, and the firmament is studded with stars.* The ballad singer makes the introduction.
13		The horseman, whose arrival the ballad singer describes, comes from afar.
14		He hails Leonore.
15–18	2, 2	Leonore appears at the window. Conversation between her and the horseman. She lets herself be persuaded to follow him, and swings off her cap.
19–20	3, 1	*The scene is a barren area with thorns and thistles.* The horseman appears in the air, Leonore with flying hair behind him on the horse. The ballad singer describes the cavalcade.
21	3, 2	The ballad singer announces a funeral train.
22		The choristers sing, "let us bury the body"; the complete funeral train appears. The horseman calls to the sexton and the priest.
23	3, 3	The coffin disappears, and the horseman pursues his course. The ballad singer narrates.
24	4, 1	*A barren area. Mountains, rocks.* A gallows, the horseman with Leonore. Narration of the ballad singer. Conversation between the horseman and Leonora.
25	4, 2	Little devils dance around on the gallows: the horseman calls them to him.
26		The rabble follows, and the journey continues in the air, which the ballad singer narrates at the same time.
27–28	5, 1	*In the foreground the theater shows a churchyard surrounded by an iron fence. Deathly: stones, coffins, countless skulls. Before the churchyard, a wild area.* The horseman with Leonore accompanied by airy rabble, the ballad singer narrates. The airy rabble disappears gradually. The horseman mocks Leonore over the fear of the dead.

TABLE 3.2. (*continued*)

STANZA	ACT, SCENE	
29	5, 2	They arrive at the gate of the churchyard, whose tombstones appear illuminated. The gate opens by itself. The horseman stops with Leonore in the middle of the churchyard. The ballad singer narrates the scene.
30–31	5, 3	The horseman falls as a skeleton from his horse and disappears: the horse throws Leonore to the ground, who, with much hand-wringing, sinks with flames into the earth. The horse disappears. Master Heine stands there with hourglass and scythe. Narration of the ballad singer.
31–32	5, 4	Leonore's tomb in the fire. Gargoyles and devil faces are the ornaments of her urn, which Master Heine places to the side. The spirits dance the dance of death in the air and sing in chorus: "patience, patience, even when your heart would break," etc.

shadow-play. Thus, on the one hand, the descriptions demonstrate the need to make concrete things left up to the imagination by the poem; on the other, they reflect ideas about what would make a good effect in shadows as distinct from in verbal description. For instance, the opening scenery features a street with "a single farmhouse," a detail Bürger does not specify. But Bürger's opening lines imply Lenore's bedroom as the scene of her initial plaints:

Lenore fuhr ums Morgenrot	Lenore stirred at first dawn red
Empor aus schweren Träumen:	From out her dreams so deep:
"Bist untreu, Wilhelm, oder tot?	"Art untrue, Wilhelm, or art dead?
Wie lange willst du säumen?"	How long your tarry keep?"

In the shadow-play, Lenore steps out of her house before speaking these lines, which serves to keep the action in the same exterior setting for stanzas 1–11.

If such changes were made mainly for logistical reasons, they could also spring from a leap of imagination. Nowhere is this clearer than in the treatment of the nocturnal horse ride, which is said to take place "in the air" (Der Reuter erscheint in den Lüften).[82] Images of Lenore and Wilhelm riding in the air are not common among visual depictions of the ballad. An illustration from Johann Rudolf Zumsteeg's musical setting of 1798, for example, shows the back legs of Wilhelm's horse firmly on the ground (figure 3.6). And indeed, when the ride begins (stanza 19), Bürger evokes the sound of the horse's gallop:

Und als sie sassen, hop hop hop!
Gings fort in sausenden Galopp

And as they sat, clop clop clop!
In speedy gallop off they went

The same sound returns in stanza 23 as they continue their hasty ride, kicking up stones and sparks as they go:

Und immer weiter, hop hop hop!
Gings fort in sausenden Galopp
Dass Ross und Reiter schnoben
Und Kies und Funken stoben.

And on and on, clop clop clop!
In speedy gallop off they went
The horse and rider snorting
And stones and sparks flying.

Bürger also describes bridges thundering under the horse's hooves as the countryside flies by (Wie flogen Anger, Heid und Land!/Wie donnerten die Brücken!; stanza 20).[83]

FIGURE 3.6. Johann Rudolf Zumsteeg, *Lenore* (Hamburg: I. A. Böhme, [1798]), title page. Jean Gray Hargrove Music Library, University of California Berkeley, M1621 .Z83 1800.

The transformation of Lenore and Wilhelm's ride into a flight through the air enhanced its supernatural appearance, making plain from the first that it was no ordinary ride. The change might be attributed to the exigencies of the shadow-play: a horse flying in the air would not need to gallop, a motion of limbs that would have been difficult to execute naturalistically with a shadow figure. Limitations in shadow-play mechanics, in this instance, made the medium more suitable to supernatural than natural representation. While perhaps a response to limitations, the change also exploits special, otherworldly potentials inherent in the shadow-play medium. Later, Romantic thinkers would make such potentials explicit. In a letter to Justinus Kerner of 1809, for example, the poet Ludwig Uhland observed:

> You can therefore drive these plays in such formations that one couldn't physically, i.e. a shadow-play can accomplish what only the imagination can follow. Optical games for the inner eye. Also, I am drawn to those plays that are not content to mimic the mechanisms of life, but rather through their magic bring forth entirely new relationships.[84]

While it predates such Romantic slights to the merely mimetic and enthusiasm for "bringing forth entirely new relationships," the "Lenore" performance can be seen to encourage such lines of thought. It exemplifies the kind of space and activity that fostered the development of such ideas—namely, an engagement with poetry and shadow-play.

And with music. For, crucially, the idea that Wilhelm's horse flies in the air gains plausibility from André's setting. When Wilhelm first arrives at Lenore's house, in stanza 13, André's accompaniment features a typical triplet galloping figure, the downbeat bass notes alternating with repeated chords in the right hand. Once Lenore joins him and they take off through the night, however, such repeated triplet chords are no longer heard. Instead, descending triplet scales become the characteristic sound of their ride. These passages occur where the text describes the countryside passing by.[85] As can be seen in example 3.4, what "flies" in these passages are the sights to the left, right, and above Lenore and Wilhelm, and it is this visual effect of rapid motion that André's music seems intended to capture. Meanwhile, the sounds of the horse's hooves, though described by Bürger, go unregistered by the piano accompaniment. The musical result is, indeed, a sense of flight through the air.

The shadow-play thus implements a shift of interpretative focus: text need not be the sole arbiter of meaning, the success of the music measured by its conformance to the poem. The combination of through-composed music and moving image allows words to recede in importance, sight and sound to take over to yield alternative meanings. The measure of success, then, may be not closeness to prior experience but the compelling revelation of "entirely new relationships."

Pedagogy of the Imagination

In 1813, Friedrich Rochlitz surveyed the musical history of Bürger's "Lenore," and with it, the genre of the ballad. Identifying Kirnberger's setting (erroneously) as the first, he described it as "cold and dry" because its strophic and syllabic approach reflected "merely the ballad—merely the genre and external form." Because Kirnberger accommodated highly contrasting stanzas to a single melody, his music was "general and expressionless"; anyone could be expected to find such repetition of the same bland music for thirty-two stanzas intolerable.[86] André then ushered in a change: Rochlitz dubbed him the "father of the concept of the ballad," in which "music is brought very close to all the particulars of the text," the "so-called through-composition" giving rise to "a kind of cantata where every strophe receives its own music."[87] Now it was not external form but "imagination and feeling" that ruled, and "what place they have once conquered . . . they will hardly give up again, but rather only expand and decorate more and more . . ."[88] Though the trend might be taken too far, the taste of the times was decidedly for "the ever-expanding grasp of the music against the claims of the poem."[89]

The decades between André's first *Lenore* and Rochlitz's review indeed saw through-composed ballads proliferate, settings of Bürger's "Lenore" included. Friedrich Ludwig Aemilius Kunzen's "Lenore" appeared in his 1788 collection *Zerstreute Compositionen für Gesang und Clavier*; Johann Rudolf Zumsteeg issued a stand-alone setting in 1798; Wenzel Johann Tomaschek's followed in 1801; and André's own revised versions date from 1782, 1788, and 1791. Yet, it is misleading to describe a linear chronological or teleological progression from simple strophic to elaborate through-composed forms, as not only Rochlitz but also modern musicologists have done.[90] Rather, strophic and through-composed coexisted from the first. But there was a shift in preference toward the latter beginning in the 1780s—a shift that decisively redefined the genre by 1802, when Koch wrote in his *Musikalisches Lexikon* that "the melody of a ballad . . . is not bound to a particular form or meter. For some time now, people have begun to throughcompose the text entirely, rather than repeat the music for each strophe of the text, as with the Lied."[91]

This shift of the tides bespeaks broad social changes whose effects rippled through the musical world, among which we might count the waning luster of populism in the wake of the French Revolution, and the steadily rising stock of composers as growing numbers of people attended concerts and purchased printed music. But a shaping force in this shift—a factor influencing its particular realization—was also the rise of moving-image apparatuses. These apparatuses challenged the established media-aesthetic equation. When Gotthold Ephraim Lessing theorized the difference between painting and poetry in 1766, he praised poetry for its ability to transcend the limits of materiality: "although the poet likewise makes us think of the goddess as a human figure, he has nevertheless removed all ideas of coarse and heavy matter and he has enlivened her body with a force which exempts it from the laws of human locomotion."[92] Visual arts, by contrast, mired the imagination in materiality, and in consequence were inferior to poetry, which allowed the imagination to work in freedom and autonomy.

The shadow-play, however, was a visual medium that—like the imagination—could picture things that defied the laws of matter, as both Berberich's horse ride "in the air" and Uhland's "optical games for the inner eye" make clear. The Regensburg performance additionally explored the relationship between shadow-play and imagination in an original verse, recited by the ballad singer as a postlude to "Lenore." Here, the ballad singer imparted a new lesson:

Und wenn Ihr ja einst lieben wolt,	And if you do want to love once,
So bleibt der Mütter Lehren hold:	Remember the mother's lessons:
Flieht alle Phantaseyen,	Flee all fantasies,
Vapeurs, Empfindeleyen,	vapors, sentimentalities,
Laßt Euren Geist in sanfter Ruh,	Preserve your spirit in gentle peace,
Durch keine Bilder schröcken.	By no pictures be frightened.
Drükt Unschuld Euch die Augen zu,	Press innocence to your eyes,
Wird Euch die Freude wecken.	Joy will awaken you.[93]

Rather than reiterating Bürger's message not to quarrel with God, Berberich's ballad singer turned to more material matters of physiology and visual culture. Counseling to "flee all fantasies" and "by no pictures be frightened," the ballad singer conflates internal and external stimuli, figuring both as threats to well-being. The lesson follows not from Lenore's experience in the ballad, but rather from spectators' experience of it. The "Lenore" performance becomes a lesson in media consumption, the shadow-play—"for pleasure as well as horror"—externalizing the overactive imagination, and thereby demonstrating both its attractions and its dangers.

We might go further still, and say that in challenging the distinction between temporal and pictorial arts, and teaching a particular conflation of imagination and visual medium, the shadow-play also fostered a mode of listening better suited to through-composed than to strophic songs. The double transformation of the ballad singer's trade—his strophic into through-composed song, and his poster into moving image—suggests such an interrelation between visual medium and mode of listening. One facet of this interrelation was the turn from singers' expressive delivery to audiovisual experiences conjured by the performance—a turn thrust upon spectators of the shadow-play. Another, perhaps more subliminal facet concerns the mental vantage point from which to apprehend events unfolding in time. As a modernized balladeer's canvas, the shadow-play not only transformed static, mute pictures into animate, "speaking" ones; it also replaced the multiple squares of the poster with one large frame. With its standard grid layout, the ballad singer's canvas provided a visual analogue to the strophic form of the ballad. Like each block on the grid, each strophe offered an identical container for different content, and these containers were to be followed in series. The frame of the shadow-play instead presented a single image that changed in time. To critics like Goethe, who preferred strophic settings, through-compositions seemed to destroy the unity and vocal expressivity of song. The framed moving image, however, supplied a perspective from which all the variety and contrast of a through-composed setting could be heard as part of one whole unfolding before the listener. By the end of his life, Goethe himself had absorbed an appreciation for through-composition directly related to his visual apprehension: in 1830, upon hearing Wilhelmine Schröder-Devrient sing Schubert's "Erlkönig," Goethe reportedly remarked: "I have heard this composition once before, when it did not appeal to me at all, but sung in this way, the whole shapes itself into a visible picture."[94]

The addition of moving images to musical performance has often been criticized for shutting down the freedom of imagination. In the late eighteenth century, however, they functioned to open up a new kind of musical engagement: they promoted imagination over judgment, immersion in another world over participation in the environment. What performance conditions—and what musical techniques—are best calibrated to foster imaginative listener engagement has remained a matter of debate ever since. The question became particularly acute at the turn of the nineteenth century in response to a new musical work: Haydn's *Creation*.

FOUR

Haydn's *Creation* as Moving Image

HAYDN'S *CREATION* PREMIERED IN VIENNA IN 1798, AND FOR nearly a year was performed only for invited, aristocratic audiences; then, in 1799, its public life began with a performance at Vienna's Burgtheater. The *Allgemeine musikalische Zeitung* reported on the remarkable "silence and attention [with which] the entire oratorio was heard—only gently interrupted at the most striking passages by soft exclamations."[1] "Even for the general taste," wrote the *Zeitung für die elegante Welt*, "there must have been a hint that extraordinary things were happening and that a work had been born which has much to give, both to the ear and to the imagination."[2] By 1801, the work had an international reputation as a "*non plus ultra* of art."[3]

Amid the enthusiasm, however, there was also ambivalence about Haydn's oratorio on the first six days of the world—an ambivalence hinted at by the condescending tone taken by the *Zeitung für die elegante Welt* toward the "general taste." Critical suspicions centered especially on Haydn's tone-painting (*Mahlerey*), the usual term for music that imitated objects. As Friedrich Wilhelm Josef von Schelling wrote in 1801, tone-painting was "something that only a debased and decadent taste can demand of music, taste of the kind that nowadays enjoys the bleating of sheep in Haydn's *Creation*."[4] Haydn's tone-paintings, it is now generally considered, made for broad appeal, while his sublime choruses satisfied the more elevated, serious demands of Romantics like August Lewald, whose distaste for presenting *The Creation* with illuminated images shall be considered at the end of this chapter.[5]

A preference for expression over tone-painting emerged in the mid-eighteenth century, as aesthetic theorists reasoned that music excelled at the former, while imitating nature demanded a kind of precision for which music was ill suited (Hiller's account of successful *ombra* music, communicating a fearful impression rather than objective qualities of a ghost, exemplifies the growing emphasis on expression; see chapter 2).[6] *The Creation*, however, prompted newly vehement reactions to tone-painting. As Richard Will has noted, the discourse surrounding the oratorio "became so heated that tone-painting acquired an even more dubious reputation than either a mimetic or a Romantic viewpoint really warranted."[7] Will identifies moral and national interests that colored discussions of tone-painting, raising the stakes on the aesthetic issue. Another concern—activated especially by Haydn's *Creation*—centered on the quality of the listening experience, and its resemblance to spectatorship of moving-image entertainments. In this concern, the reception of Haydn's *Creation* intersected with cultural debates taking place across Europe. In his study of early nineteenth-century British culture, Gillen D'Arcy Wood—with a nod to M. H. Abrams's account of mimetic versus expressive theories of art in *The Mirror and The Lamp*—identifies a conflict "between the lamp and the magic lantern: between Romantic, expressive theories of artistic production emphasizing original genius and the idealizing imagination, and a new visual-cultural industry of mass reproduction, spectacle, and simulation."[8] While the industrialization of moving-image entertainments remained decades off, *The Creation* inspired theories of both creative genius and simulative spectacle.[9] The relationship between the two, far from being antagonistic *a priori*, was worked out through the specific qualities (aesthetic, social, and epistemological) of their exemplary instances—namely Haydn's music, the magic lantern, and the shadow-play.

The entanglements of *The Creation* with contemporary moving-image technologies can be seen most clearly in the writings of two critics: Johann Karl Friedrich Triest and Carl Friedrich Zelter. Whereas Zelter joined the enthusiastic concertgoers who assured that *The Creation* would become one of the most frequently performed works of the early nineteenth century, Triest sided with those who censured the work. Yet, both turned to image-projection apparatuses to articulate their views:

> And what can aesthetics have to say to a natural history, or geogony, set to music, where objects pass before us as in a magic lantern [*magischen Laterne*]?
>
> –Johann Karl Friedrich Triest, 1801

> Movement and repose are made to come alive by a magical play of color on the imagination and by the art of the music; all this is paraded in front of the inner eye like a fine shadow-play [*hoheres Schattenspiel*],

> showing us the beauties of paradise, a wonderful garden, or a world newly born.
>
> –Carl Friedrich Zelter, 1802[10]

In late eighteenth-century Germany, magic-lantern shows and shadow-plays were performed by different kinds of showmen in different social spaces. But both belonged to the realm of "natural magic," as popular scientific literature in Germany typically called its subject matter. More clearly than its French counterpart "rational recreations," the term "natural magic" foregrounded the problematic resemblance between technological wonders and supernatural events. As Johann Samuel Traugott Gehler explained in his *Physikalisches Wörterbuch* [Physical Dictionary, 1790], "magic in general is the art of producing results that seem to surpass the natural powers of the body. One thus divides this art into natural magic, where the seemingly wonderful achievements can nevertheless be explained by forces and laws of the body, and into the supernatural, which would require the involvement of the spirits."[11] Pairing seemingly superhuman achievement with represented content, moving images were technological wonders as much as storytelling media.[12] While such spectacles had educational potential as optical-mechanical demonstrations, they also had a more strictly sensory appeal, and in the eyes of critics they all too often merely astonished the ignorant, or worse, promoted belief in the supernatural. For turn-of-the-century spectators, moving-image entertainments promised to delight, offered to enlighten, but also threatened to stupefy or deceive.

At once descriptive and evaluative, the magic lantern and shadow-play evoked the audiovisual experience of listening to Haydn's *Creation* while also serving as rhetorical instruments in a critical debate. Triest's and Zelter's metaphors thus point beyond the reasoning of aesthetic theorists to the resonances of contemporary audiovisual culture, and the divergent reactions made possible by the oratorio's resemblance to technological spectacles. Haydn's *Creation* crystallized a formative moment for musical culture not only with respect to the status of the composer, but also in its stance toward visual technologies. The moment was formative too for these technologies, as the permeable boundary between philosophy and spectacle hardened to separate knowledge-producing scientists from profit-seeking showmen. The debates over whether *The Creation* constituted vulgar entertainment or a great work of art thus also entailed questions about the value of moving-image presentations, the status of seemingly supernatural powers, and the proper modes of musical listening.

Magic Lanterns and Barrel Organs

The magic lantern was a device for projecting images drawn on glass slides, its name evoking the phantom quality of the projected images as well as the mysterious means by which these images were produced. Like the peepbox, the magic lantern emerged as an instrument of natural magic in the seventeenth century, embraced especially by Jesuits. First mentioned in optical writings of the 1650s, the instrument became generally known to natural philosophers by the 1670s, and by century's end "even the least able glass-grinder" was considered capable of producing and arranging its components.[13] The design of a magic lantern was indeed quite simple (figure 4.1). A candle or wick lamp produced light inside a metal lantern-shaped container. On one side of the container, an opening allowed this light to shine through a tube, at the end of which was a lens. At the base of the tube, one inserted a glass slide onto which images had been painted with translucent paints. The light, shining through the glass slide and focused by the lens, projected the painted images onto a nearby surface, usually a sheet hung on a wall.

The magic lantern was sibling to the peepshow, the two being so closely associated that, as we saw in chapter 2, their names were often used interchangeably despite clear differences between the devices. Like the peep-

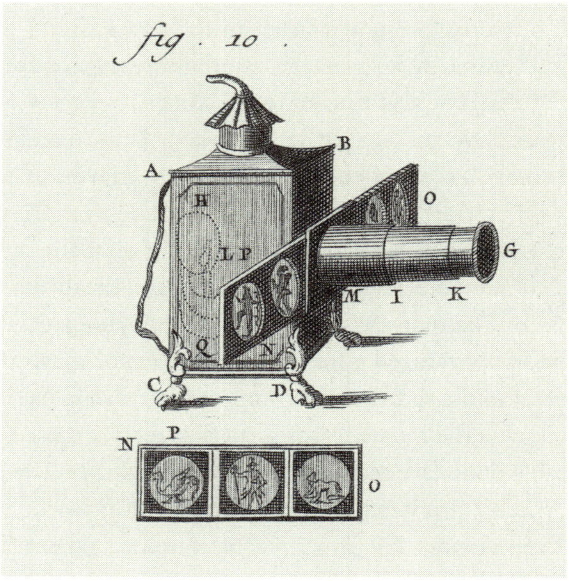

FIGURE 4.1. Detail from "Optique" plate, *Recueil de Planches, sur les Sciences, Les Arts Libéraux, et Les Arts Méchaniques, avec leur Explication*, vol. 5 (Paris, 1767). Kislak Center for Special Collections, Rare Books and Manuscript, University of Pennsylvania Libraries.

box, the magic lantern occupied two distinct social spaces. On the pages of eighteenth-century encyclopedias, optical treatises, and popular science books, magic lanterns appeared as rational instruments and commodities for purchase from instrument makers by upper- and middle-class citizens. On streets and fairgrounds, meanwhile, itinerant Savoyards carried magic lanterns from place to place, offering with them a cheap entertainment for anyone who would pay a few pennies. The juncture of these two realms—the elite and literate with the lowly and oral—is illustrated by an anecdote from 1737, when Voltaire used the optical instrument to entertain a gathering of friends at the chateau of the Marquise du Châtelet at Cirey-sur-Blaise. As one guest recalled, "after supper, he gave us the magic lantern, with some remarks which would make you die laughing . . . all kinds of tales always with the Savoyard accent."[14] Like the peepshowman, the magic lanternist provided an attractive object of imitation and parody, particularly for the way he narrated his series of images.

While Savoyards presented peep- and magic-lantern shows with similar verbal commentary—labeling the sights with frequent injunctions to look, in their strong foreign accent—the logistics of exhibiting the two differed substantially. Whereas the peepbox created an enclosed space wherein sun- or candlelight illuminated the sights, the magic lantern required a dark environment for its projected images to be visible. So, while peepshows were exhibited outside in broad daylight, magic-lantern shows usually took place in the evening and indoors, the exhibitor being invited into the home to present his show. Here, he would also likely display a mechanical dancing doll named Margaretha or Catherina, as featured in the scene reproduced in figure 4.2. The magic lanternist, still carrying his device on his back, appears at the right of the scene, which is set in a suitably dark interior.[15]

Completing the magic lanternist's technological ensemble was the hurdy-gurdy or, more often, the barrel organ, as can also be seen in figure 4.2. We saw in chapter 2 that peepshowmen used these instruments to help announce their shows, their tones being one of the sonic tools with which they aroused the desire to look; Wölfl's setting of the peepshow in *Das schöne Milchmädchen* additionally suggested the use of the barrel organ in alternation with verbal narration, to signal a new picture. Depictions of magic lanternists likewise suggest both advertising and presentation functions for the barrel organ, with the latter becoming possible in the latter half of the eighteenth century. That is, magic lanternists seem at first to have played the instrument while roaming the streets to help announce their show: the lone Savoyard commonly depicted in early to mid-eighteenth-century

FIGURE 4.2. "Le charmante Catin," engraved by Louis-Madeleine Cochin (after 1740, before 1767). Bibliothèque Nationale de France.

prints and porcelains—playing a barrel organ while carrying a magic lantern on his back—could not have provided organ accompaniment to his magic-lantern show, when his hands would have been occupied shifting the slides. Around the 1760s, however, magic lanternists began to appear in the company of a musical companion, who played not only as they wandered together but also during the magic-lantern show. The shift to teamwork suggests that the magic-lantern show became both more elaborate and more lucrative, its exhibition sustaining the efforts of more than one showman. Depictions from the late eighteenth and early nineteenth centuries commonly show one Savoyard aiming the magic lantern at the wall while another holds the barrel organ, his hand on the handle, cranking or ready to crank out a tune (figure 4.3).

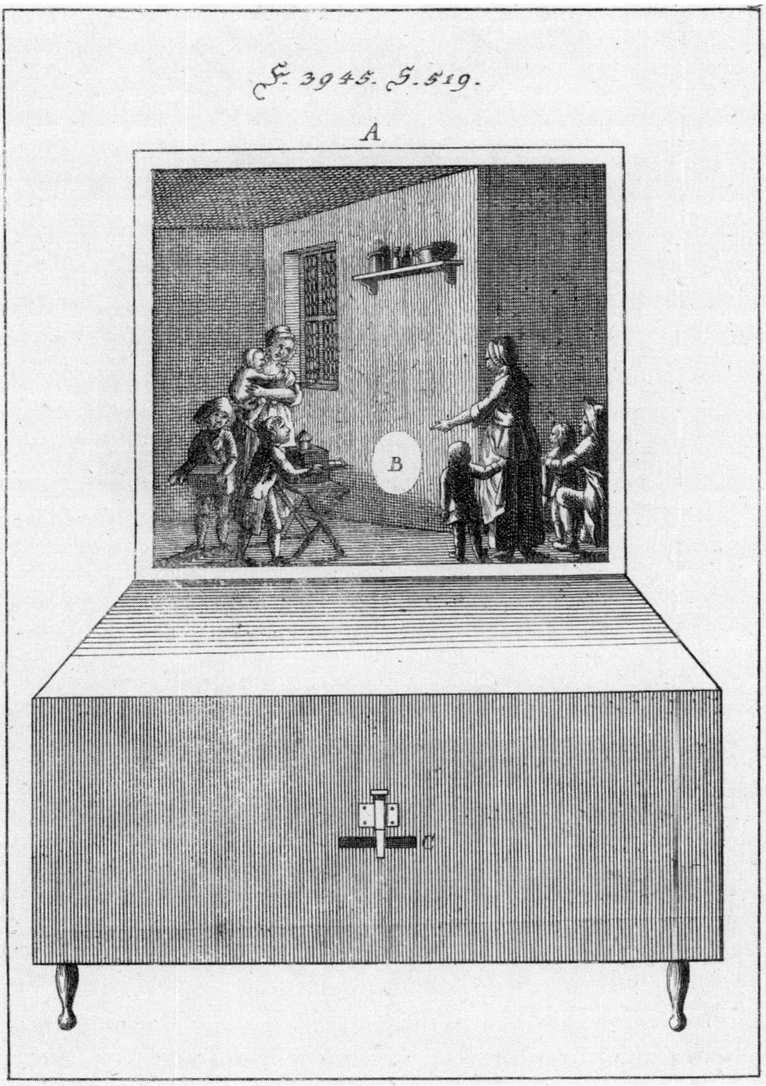

FIGURE 4.3. "Mechanische Zauber-Laterne" from Johann Georg Krünitz, *Oeconomische Encyclopädie, oder, Allgemeines System der Staats- Stadt- Haus- und Landwirthschaft*, vol. 65 (Berlin: Joachim Pauli, 1794). The Bancroft Library, University of California-Berkeley, AE27.K7 v. 65.

FIGURE 4.4. Magic lanternist and barrel organist from Mathew Deisch, *Danziger Ausrufer* (Danzig, 1763–1789). Reproduced from Max Herrmann, *Jahrmarktsfest zu Plundersweilern* (Berlin: Weidmannsche, 1900).

What kind of tune is suggested by Mathew Deisch's street-crier series *Danziger Ausrufer* (Danzig, 1763–1789), a rarity in the genre for presenting not only the characteristic words but also the melodies of the street-criers it depicts. Here, a Savoyard carrying a magic lantern on his back advertises the dancing doll "charming Catherine," while his organ-cranking companion plays a minuet and cries, "Laterna magika, Zattenspiehl an der Wandt" [Magic lantern, shadow-play on the wall] (figure 4.4).[16] The minuet is likely a transcription of street barrel-organ music of the period: the treble register matches the small size of the instrument in the accompanying illustration (too small to accommodate the length of pipe required for bass notes), while the two-voice texture and selection of pitches are consistent with the

description of street organs in Dom Bedos's authoritative treatise on organ manufacture, *L'Art du facteur d'orgues* (1766). The melody of the minuet is also characteristically mechanical: dominated by arpeggios and other leaps in running eighth notes, it is the kind of music suited less to the voice than to the fingers—especially to the automated "fingers" of a pinned barrel.

The history of the street barrel organ has focused on its roles as a noise polluter and attention destroyer, its cranked melodies contributing to the increasingly machine-made cacophony of the urban soundscape and to the misery of intellectual workers attempting to concentrate.[17] The street organs of the eighteenth century seldom provoked such lines of complaint, however. Significantly smaller and quieter than their nineteenth-century descendants, eighteenth-century street barrel organs instead lent a gentle fluting to their environment. By the early nineteenth century, they represented a mechanical music of almost comforting purity. As one of E. T. A. Hoffmann's characters, contrasting street organs with musical automata that imitated human form, explained, "I infinitely prefer the commonest barrel organ, in which the mechanism attempts nothing but to be mechanical, to Vaucanson's flute player, or the harmonica girl."[18] Similarly, Hegel maintained that performers—if they were to be artists—must not "sink to being merely mechanical, which only barrel-organ players are allowed to be."[19] Significantly, Hegel found not just the instrument or music to be mechanical, but also the player—the Savoyard. (This notion found its logical extension toward the end of the nineteenth century, when the German company Polyphon manufactured a line of coin-operated, barrel-organ-playing Savoyard automata.)

Hoffmann's and Hegel's remarks raise questions about the status of musical machines in the eighteenth and early nineteenth centuries. A number of scholars have placed musical machines at the heart of a philosophical upheaval in the eighteenth century, centered on the replacement of human functions by manufactured equivalents; the implication—agreeable to some, troubling to many more—was that such precious human endowments as emotion, intelligence, and the soul were ultimately reducible to mere material mechanism. From this perspective, musical machines represented existential threats to human freedom, subjectivity, and the immortal soul. Other scholars, however, have sought to revise our understanding of the reception of musical machines in the period. Adelheid Voskuhl has demonstrated an absence of anxiety over musical automata in the eighteenth century, showing that court and artisan culture, rather than philosophical questions or the first murmurings of industrialization, account for their manufacture and reception.[20] Terrance Riley argues that the whole range of mechanical musical instruments and automata were based on long-known clockwork, and that by the turn of the nineteenth century familiarity had bred complacency. Riley uses two early nineteenth-century devices (the Panharmonicon and the Componium) to demonstrate "an ease with human-machine mediation and a technical skill at recreating human intellectual functioning," concluding that rather than being philosophically charged, musical machines were made for "a more homely and domestic fascination with mechanized music."[21] Focusing on the case of orchestra-imitating machines, Emily Dolan finds similarly that through the early nineteenth century, "the relationship between the human and the mechanical was not yet one of opposition."[22]

If the human and the mechanical were not yet in opposition, however, there were oppositions between human and human—between enlightened and vulgar, cultivated and primitive, native and foreign—that factored into the status of musical machines. Rather than make blanket claims for musical machines in the period, then, one can look at how these (like magic lanterns and peepshows) occupied multiple social spaces, their value and significance changing with how they were used and by whom. Haydn could write music for royal flute clocks, and Mozart for a mechanical organ heard hourly at a public waxworks, without artistic crisis, even as similar mechanisms heard on the street stood beneath their compositional dignity.[23] In his brief *Historical Treatise on the Organ* (1756), the North German music director Johann Gottfried Mittag observed that barrel organs "have no particular use, hence they come to no esteem. They are heard often in the streets from wandering Savoyards, which at the same time announce that they will present the magic lantern or bella Margaretha; but they belong in the category of trivia."[24] Since barrel organs were also to be found modeling songs for pet birds and providing music at provincial churches, it was not truly a lack of uses that brought them to no esteem. Rather, it was the particular sphere of use that Mittag identified: that of itinerant entertainers. These itinerants were hardly to be distinguished—in terms of social standing—from the beggars initially identified with the instrument: already in the 1720s, an organological treatise placed the portable barrel organ in the hands of foreigners seeking alms, the pins of the rotating barrel replacing the movements of the fingers.[25] Far from shaking a fundamental faith in the human, the barrel organ merged seamlessly with its poor, wandering operator, whose feeling and intelligence were already in doubt.

As the magic lantern became a familiar evening's entertainment, so too did its literary appearances proliferate. Like the peepshow, it often figured as a trivial entertainment for children. But in contrast to the discursive uses of the peepshow, which tended to center on the arrangement of peeping, those for the magic lantern centered on the fleeting, apparitional quality of its images. As a metaphor, the magic lantern typically figured rapidly passing sights (either in the world or in the mind), or the chimerical status of a perception or claim. In the late eighteenth century, the magic lantern evoked such notions with a range of tones and valences. Goethe, for example, invoked the magic lantern wistfully in *The Sorrows of Young Werther*, using it to figure perceptions colored by love: "what is the world to our hearts without love? What a magic-lantern is without light! You have but to insert the little lamp, and the brightest figures appear to you on your white wall! And, if they are nothing but fleeting phantoms, they yet make us happy when we, like mere children, stand before them and are delighted by the miraculous apparitions."[26] For Kant, by contrast, the illusory status of magic-lantern images made them insidious. Seeking to discredit proponents of uncritical philosophies, Kant cast them as illusionists who, "as through a magic lantern, make miraculous things momentarily manifest, only thereafter swiftly to vanish, leaving behind still among the naive a sense of wonderment and the impression that something extraordinary must be behind it all, which they just didn't catch."[27] In such discursive contexts, the magic lantern was not only cheap and lowly like the peepshow, but also insidiously deceptive. The feelings of astonishment and wonder felt by its spectators suggested not just childish naïveté but a dangerous ignorance—a susceptibility to belief in false appearances.

When Triest characterized *The Creation* as a magic lantern, the grounds of similarity went beyond the production of fleeting, apparitional images. What made *The Creation* like a magic lantern—what made the metaphor so effective a means to belittle the work—was the very form and substance of the oratorio's storytelling. In fact, one of the most popular stories told with the magic lantern was that of the creation of the world.

Magic-Lantern Audiovisuality

The Bible was an important source of subject matter for both peep- and magic-lantern shows, but the creation story was especially associated with the latter. No small part of its attraction lay in the dramatic potential inherent in God's creation of light: with the audience first in darkness, lighting the lantern's lamp provided a powerful means to represent this all-important event. At a time when spectatorship in darkness was still unusual, moreover, the creation story turned darkness into part of the performance, giving narrative meaning to an unfamiliar, and potentially uncomfortable, social situation.

The traces of magic lanternists' oral performances survive in various kinds of texts. In chapter 2, we saw that Savoyard peepshowmen related vocals to the visual in two distinct ways: to advertise and lure viewers to their show, they bawled a street cry that characteristically lapsed from words into vocables such as "hahiha"; to accompany their images, they provided verbal commentary that consisted mainly of naming the sights shown. Magic lanternists also employed these two vocal modes but configured them differently within their shows, to fit the conditions of communal spectatorship and the logistics of changing slides. An item from an early nineteenth-century collection of dialects, titled "Schattenspiel-Lied eines Savojarden, von Adam und Eva" [Shadow-play song of a Savoyard, on Adam and Eve], exemplifies the common pattern for magic-lantern performance.[28] The Savoyard introduces his entertainment and narrates its pictures, the text representing his accent phonetically. But rather than lapsing into "hiha" vocables, the magic lanternist—who was more consistently identified with the barrel organ—punctuates his recitation with vocal imitations of his mechanical instrument. Here, the imitation takes the form of "diedel—diedel—diedeldum," followed by "Pretty shadow-plays!" The lines resemble a street cry but function as a refrain, providing a recurring interlude in the commentary addressed to the audience:

Woll si schäne Schattenspiel?	You want pretty shadow-play?
Hab sie Spaß, und darf nit viel	You have fun, and it doesn't take much
Mir dafür bezahlen.	For me thereby to enchant.
Ick will euck hier an die Wand	I will here on the wall
Durck di Schatten, allerhand,	Through the shadows, of all sorts,
Kanz posirlig mahlen.	Paint everything possible.
Dreh dazu die Orgel um:	Crank the organ to them:
Diedel—diedel—diedeldum—	Diedel—diedel—diedeldum—
Schöne Schattenspiele!	Pretty shadow-plays!
Guck si her! Was seh' si? Nix!	Look here! What do you see? Nothing!
Und ick explicire six:	And I explain sights
Daß si alles sehen;	That you all see;
Himmel, Erde, Luft und Meer	Heaven, earth, air and sea
Alle Dinge, ließ der Herr	All things, God let
Aus die Nix entstehen.	form out of nothing
Nix ist kroß Miraculum!	Nothing is great miracle!
Diedel—diedel—diedeldum—	Diedel—diedel—diedeldum—
Schöne Schattenspiele!	Pretty shadow-plays!
Erstlick, will ick präßentir:	First, I want to present:
Adam und die Eva hier	Adam and Eve here
In die Paradißel;	in paradise;
Wie die erste Menchen paar,	How the first human pair
In die Stand von Unschuld war	Stood without shame

Munter wie die Wiesel;	Jolly as the weasel;
Haben Freude um und um.	Joy around and around.
Diedel—diedel—diedeldum—	Diedel—diedel—diedeldum—
Schöne Schattenspiele!	Pretty shadow-plays! . . .[29]

On the page, the barrel-organ refrain appears to provide closure, coming at the end of each visual segment. However, recalling the use of the organ to announce each new picture in the peepshow scene of Wölfl's *Das schöne Milchmädchen*, as well as the usual preparatory function of the street cry, its seems likely that in performance the barrel-organ refrain would have generated anticipation, preceding each new series of pictures.

A similar magic-lantern performance appears in Goethe's *Jahrmarktsfest zu Plundersweilern*, a farcical theater piece penned in 1773 and first performed in 1778 before an invitation-only audience at Duchess Anna Amalia's summer palace, Schloss Ettersburg outside Weimar. The play features a series of street-criers offering their wares at an annual fair. As evening sets in, a *Schattenspielmann* [shadow-play man] is heard from behind the stage crying, "Orgelum Orgeley! Dudeldumdey!" This loosely onomatopoetic phrase was connotatively loaded. By 1770, "dudeldum" and "dudeldei" were established terms for music or poetry having no meaning or value, the association being with a drone.[30] "Orgelei," at this point, served to imitate the tone of a barrel organ, but by the early nineteenth century, it had additionally become a more general term, meaning tedious or unctuous speech.[31] Savoyard entertainers seem to have provided the origin for this pejorative usage. Gottfried Bürger kept the root source alive when in 1781 he likened a play with poor music to that "old Savoyard-style Orgelum orgelei, wherein word and melody are equally bad."[32] Though they too sang songs, Savoyards were not members of the German folk Bürger so eagerly mined for poetic material, as discussed in the previous chapter.

Two fairgoers, a doctor and bailiff, immediately recognize "Orgelum Orgeley! Dudeldumdey!" as the signal for a magic-lantern show. The former remarks, "let him come . . . Put out the lights/We are in an honest house," alluding to period worries about unsavory or criminal activity in darkened performance spaces. The latter directs, "bring out the sheet."[33] The Schattenspielmann then enters and begins his performance in the dark, reserving the illumination of the lamp for the creation of light. As in the "Schattenspiel-Lied eines Savojarden," the vocal imitation of the barrel organ provides a refrain:

Orgelum, orgeley!	Orgelum, orgeley!
Dudeldumdey!	Dudeldumdey!
Ach wie sie is alles dunkel	Oh, how everything is dark,
Finsternis is	Blackness reigns
War sie all wüst und leer	All was waste and blank
Hab sie all nicks auf die Erd gesehn	Nothing to be seen on earth
Orgelum, orgeley!	Orgelum, orgeley!
Dudeldumdey!	Dudeldumdey!
Sprach sie Gott, 's werb' Licht!	Then God said: let there be light!
Wie's hell da 'reinbricht!	And brightness broke out everywhere!
Wie sie all' durk einander gehn!	How they all mix with one another!
Die Element alle vier	All four elements
In sechs Tag' alles gemacht is,	In six days, everything was made
Sonn,' Mond, Stern, Baum und Their	Sun, moon, star, tree, and animal
Orgelum, orgeley!	Orgelum, orgeley!
Dudeldumdey!	Dudeldumdey!
Steh' sie Adam in die Paradies,	Behold Adam in paradise
Steh' sie Eva, hat sie die Schlang verführt.	Behold Eve, seduced by the snake.
Nausgejagt,	Expelled,
Mit Dorn und Disteln	With thorns and briars
Geburtsschmerzen geplagt.	Plagued with birth pangs
O weh!	O woe!
Orgelum . . .	Orgelum . . .[34]

The Schattenspielmann goes on to narrate man's sinful dalliance in procreative activity, the corrective Flood, and the winged Mercury bringing an end to the horror—always returning to the refrain "Orgelum, orgeley, Dudeldumdey!"[35]

A surviving late eighteenth-century slide provides a sense of the visual counterpart to such oral performances (plate 4). The slide comes from the London merchant John Scott. A specialist in watercolor materials, Scott advertised the many virtues of the magic lantern to well-to-do customers: "a pleasing Family Amusement, well suited to all ages and sexes, ready in a few minutes for the entertainment of friends or family—in fact a chearful [*sic*] house should never be without one."[36] The slide may differ in its technical and artistic details from those carried by traveling showmen, few of which survive. Its correspondences with the narrations just cited, however, are suggestive. The slide shows the sun, then the moon surrounded by stars, followed by Adam and Eve sitting in paradise in the company of a rabbit, peacock, and lion, then Adam and Eve eating from the Tree of Knowledge, and finally their expulsion from Paradise. A presenter would have displayed these five images in succession, one at a time, by maneuvering the slide incrementally through the lantern. The depicted objects, in consequence, would have "passed before" the viewers, as writers (including Triest) typically characterized the behavior of magic-lantern images.

Most eighteenth-century slides contained three to five separate images. Goethe's Schattenspielmann seems to describe this same number of images between refrains, each one or two lines of text corresponding to an image. Putting slides and oral performance together, we can surmise that the verbal commentary coordinated with the projected images and the "barrel organ" refrain with their absence when the slide was changed. While moving between the images on a single slide could be done nearly instantaneously, switching between one slide and the next produced an interval during which no picture appeared on the wall.[37] Occurring between the slides, the magic lanternist's refrain filled the interludes in the flow of images. To put it another way, images were paired with their labels, visual absence with verbal absence; an aesthetic of redundancy governed the disposition of image, word, and "wordless" music.

We can learn more about magic-lantern audiovisuality—or rather, about how it was perceived by cultural elites—by turning to the music written for Goethe's *Jahrmarktsfest*. At its 1778 performance, the play incorporated musical numbers composed by the Duchess Amalia herself.[38] These numbers included the vendors' street cries, street singers' performances, and the Schattenspielmann's refrain "Orgelum, orgeley, dudeldumdey!"[39]

Amalia's setting of "Orgelum, orgeley" gives a musical portrayal of the magic lanternist that emphasizes simplicity, stasis, and repetition. Like the barrel-organ tune recorded in Deisch's street-crier print, Amalia's "Orgelum, orgeley" number is a species of minuet. Surprisingly, however, the instrument her music suggests is the hurdy-gurdy—that other Savoyard instrument that similarly required cranking but came without preprogrammed tunes, instead relying on the player's musical abilities. Thus, in place of extended arpeggios in running eighth notes, one finds arpeggiated triads that fall easily under the hands; a three-note-turn figure similar to the one imitating a hurdy-gurdy at the end of Neefe's "He! Raritäten" (see chapter 2); a drone (mm. 11–14); and a concluding upward melodic

EXAMPLE 4.1. Anna Amalia, *Jahrmarktsfest zu Plundersweilern*, "Orgelum, orgeley (no. 1)"

tick characteristic of hurdy-gurdy imitations (m. 18). Employing a limited harmonic language and short phrases, Amalia's "Orgelum, orgeley" is also simpler and more repetitive than the barrel-organ music in Deisch's print: the music alternates exclusively between tonic and dominant harmonies, and frequently cadences on the tonic (every four, or sometimes two, measures) (example 4.1). The number additionally features abundant repetition of short one- to four-measure units. In fact, the entire eighteen-measure piece consists of only five measures' worth of material, repeated in various groupings. This economical phrase design, atypical of Amalia's music, seems calculated to produce monotony and evoke mechanical repetition—characteristics amplified by the recurrence of the music with each "orgelum" refrain in the showman's speech.[40] Such were the musical qualities of the magic-lantern show suggested by Goethe's text and vividly captured by Amalia's music.

Haydn's *Creation* as Magic Lantern

To experience Haydn's *Creation* as "a natural history, or geogony, set to music, where objects pass before us as in a magic lantern," then, would be to perceive not a continuous series of passing shadows but rather an intermittent stream of images synchronous with their verbal labeling, the interstices being filled with repetitious, mechanistic music. The accompanied recitative "And God made the firmament" is particularly susceptible to such a hearing, and as the first of those numbers aptly described as "a natural history or geogony set to music," it may have fixed indelibly in Triest's mind the resemblance between oratorio and magic-lantern show.[41]

The text throughout *The Creation* mingles biblical passages from the book of Genesis with poetic elaborations of the same subject matter derived from John Milton's *Paradise Lost*. In "And God made the firmament," the text consists of the biblical account of God's separation of the waters above and below the firmament, followed by poetry describing a series of meteorological events. After recitativo *secco* for the biblical introduction, the number unfolds as a series of tone-paintings in the orchestra, each lasting two to four measures, in alternation with the unaccompanied voice of Raphael describing the weather phenomena one by one (example 4.2).

Haydn's tone-paintings of the meteorological happenings precede their identification in the text, an ordering also found in the accompanied recitatives "In splendor bright" and "Straight opening her fertile womb." As Nicholas Temperley has observed, there is no clear precedent for this feature of *The Creation*.[42] Rather, the assumed dependence of music on words for meaning made the contrary practice—the prefacing of musical elaboration with verbal specification—standard practice in eighteenth-century music. Neefe's setting of the peepshow in the finale to *Amors Guckkasten* provides a clear example of the text-then-music practice, where—as discussed in chapter 2—the alternation of vocal and instrumental phrases (the latter wordlessly repeating and embellishing the former) encourages listeners to "see" in the music what has just been described.

Haydn's music-then-text practice daringly prioritizes music over words within the expressive economy of the vocal genre. But it also risks confusing and disorienting the listener who does not expect events to unfold in this order. In this confusion and disorientation, the practice opens up the possibility of a magic-lantern-like experience: that of meaningless, mechanical music alternating with verbal description, the correlating visual experience being an alternating absence and presence of images. Indeed, heard as fleeting moments of clarity within an ongoing series of verbal-musical events, the brief snippets of recitative in "And God made the firmament" recall descriptions like Kant's of the magic lantern making things "momentarily manifest, only thereafter swiftly to vanish." Meanwhile, hearing the tone-paintings without knowledge of their representational targets brings their mechanical quality to the fore: they feature motoric rhythms, promi-

EXAMPLE 4.2. Haydn, *The Creation*, no. 3, "And God made the firmament"

nently placed arpeggios, and small-scale repetition, most being made up of a repeated two-measure unit. The tone-paintings thus recall both the mechanical style of barrel-organ music and the micro-repetitions of Amalia's "Orgelum"—the musical qualities of the Savoyard and his cheap shows that amaze the ignorant.

Haydn as Savoyard

When Haydn composed *The Creation*, he was already considered the leading composer of his day—a role of newfound celebrity thanks to the growing consumer market for music. Further fueling public admiration, the oratorio prompted early listeners to deploy the metaphor of creation in ways that

EXAMPLE 4.2. (*continued*)

elevated Haydn's music from representation to enactment, and Haydn from original genius to God-like creator. As one listener put it in verse: "Lately *thy* creative *fiat*!/Made thunder through the kettledrums/And heaven, sun, moon and earth/The Creation complete for a second time."[43] Nicholas Mathew has noted that Haydn's setting of "and there was light" "encouraged such rhetoric precisely by striving to erase the distinction between representational distance and physical presence, performing the creation of

EXAMPLE 4.2. (*continued*)

light with its dazzling brass and timpani and sudden switch from C minor to C major."[44] The magic lantern would seem poised to support a similar conflation of representation and enactment. The ability of the magic lanternist to conjure things into existence was central to the "magic" appellation of his device and its capacity to astonish. Especially when presenting the birth of the world—heaven, sun, moon, and earth—the magic lanternist might be thought to possess God-like powers.

EXAMPLE 4.2. (*continued*)

The Savoyard, however, stood for the very opposite of creative activity. When Johann Gottfried Herder wished to disparage the dramatist who strictly adhered to traditional rules, he called him "a wretched master of ceremonies! A Savoyard of the theater, not a creator, poet, or god of the drama!"[45] That magic lanternists merely imitated—repeating without understanding—became a common theme. Recalling a childhood encounter with a travelling Savoyard, a certain Fr. Von Beaumont reported asking the man why he turned out the lights before exhibiting his magic lantern. The showman responded that his father had thus earned his bread, and so he did the same. "Who would believe it!" remarks Beaumont, shocked at the display of ignorance, "they learned to set their machines in motion; one can ask nothing more of them."[46] In Jean-Pierre Claris de Florian's popular *Fables de Florian* (1792), one of the tales recounts the story of a magic lanternist's monkey who decides to exhibit the magic lantern himself. He assembles an audience of animals and presents the show in imitation of his master:

Messieurs, vous voyez le soleil,	Gentlemen, behold the sun,
Ses rayons et toute sa gloire.	Its rays and all its glory.
Voici présentement la lune; et puis l'histoire	Here now the moon, and then the story
D'Adam, d'Eve et des animaux . . .	Of Adam, of Eve and the animals . . .
Voyez, messieurs, comme ils sont beaux!	Behold, gentlemen, how beautiful they are!
Voyez la naissance du monde . . .	Behold the birth of the world . . .[47]

The punch line reveals that the audience sits in total darkness, seeing nothing: though the monkey puts the slide in the lantern and apes his master's oratory, he neglects the crucial act of lighting the lantern.

The Savoyard's mindless imitation was perhaps nowhere more apparent than in his music. Amalia captured this aspect of the Schattenspielmann in her setting of "Orgelum, orgeley." The relationship between instrumental accompaniment and vocal line in "Orgelum" differs from that in her other numbers for *Jahrmarktfest*: only in "Orgelum" does the singer repeat exactly the melody introduced by the orchestra.[48] Also unique to "Orgelum" is the continuity between introductory phrase and singer's melody. Whereas in every other number a full cadence prepares the singer's entrance and the singer initiates a new phrase, in "Orgelum" the singer enters in the midst of an ongoing phrase. The effect is to make the voice a needless add-on to the mechanically proceeding music. The recombinatorial aspect of the number, which is pieced together from only five measures of music, also suggests the Savoyard's limited capacities. Herder's "Savoyard of the theater" was guilty not only of slavishly following rules, but also of assembling rather than creating: "how carefully I have patched and stitched it all together!" Herder's deficient dramatist cries.

This image of the Savoyard as mere operator and assembler resonates with Triest's assessment of Haydn's work on *The Creation*, strikingly at odds as it is with celebrations of the composer as God-like creator. As the composer of the magic-lantern-like oratorio, Haydn could not do otherwise than produce the music dictated by his text: "the text was certainly not his own," Triest explained, "and it was thus not his fault that it forced him to perpetual representation of objects instead of subjects."[49] "And God made the firmament" plays into the perception of mechanical procedure, as the regular alternation between vocal and instrumental phrases takes on an automatic character. The succession of tone-paintings thus becomes the repetition of a barrel organ, predetermined by an agency outside the composer's control.

Perhaps even more disturbing to Triest than the absence of Haydn's authorial agency from *The Creation*, however, was the sacrifice of freedom the oratorio demanded in turn from its listeners. For Triest, the "actual purpose" of "purely instrumental music" was "to assist the participant, by means of the *indefinite* quality of its phrases . . . to make *freer* use of his imagination, and then to leave it up to him (or the circumstances) to determine which *more specific* ideas and feelings this play of tones may lead him to."[50] It was precisely such a process of "free listening" that Haydn seemed to initiate but then snuff out by placing the relevant text *after* his tone-paintings. Rather than allowing listeners to determine their own specific ideas and feelings (as did Haydn's symphonies and string quartets, which Triest considered the pinnacle of instrumental music), the text of *The Creation* dictated to listeners what they should imagine, just as it dictated to Haydn what he must compose.

If listeners' visual and mechanical associations with the oratorio pointed to Haydn as Savoyard, however, they could also position him as maker. For where machines could suggest human submission to external forces, they could also reflect human mastery over the natural world; and where passing images could suggest trivial entertainments for the ignorant, they could also suggest philosophical entertainments for the educated. Such competing interpretations of the intersection of mechanism and spectacle were being negotiated across a burgeoning spectrum of publications and performances, as natural philosophers and entrepreneurial showmen sought to distinguish their offerings from mindless diversions, even as they used instruments of vulgar origin or association. The English instrument maker, scientific lecturer, and author of popularizing treatises Benjamin Martin articulated the challenge in the case of the magic lantern:

> because this Instrument is capable of exhibiting a large and surprizing View of any odd, ridiculous, or monstrous transparent Picture, and thereby occasion Wonder and Surprize to an inconsiderate Populace, it does not by any Means follow that Ladies and Gentleman, nay Scholars, and Sages themselves may not be very rationally amused with the curious and noble Effects which such an Instrument can produce with proper Objects applied to it . . . There is nothing in the Magic Lanthorn that is scandalous, but the Name itself . . .[51]

The technology, Martin argued, must be judged independently of its abuse and misunderstanding by the vulgar. Martin sought to make clear the intellectual nature of the magic lantern by changing its name, substituting for "magic"—which called to mind ignorant belief in conjurors' frauds—the more technical, Greek-derived "Megalographica," meaning "producing a very large and magnified Picture of a Small Object."[52]

Similarly, Zelter sought to defend Haydn's *Creation* via a telling shift in metaphor: from magic lantern to shadow-play. Published in the same journal as Triest's article one year later, Zelter's review of *The Creation* recast Triest's characterization of the oratorio almost phrase for phrase: where Triest heard "a natural history or geogony," Zelter heard "an historical-cum-poetical description"; where Triest saw "objects pass before us," Zelter saw "paraded in front of the inner eye . . . the beauties of paradise, a wonderful garden, or a world newly born"; and where Triest compared the work to a magic lantern, Zelter compared it to a shadow-play. By invoking the shadow-play, Zelter acknowledged a similar visuality to the work—one reminiscent less of nature than of its imitation by moving-image technologies—but distanced it from cheap entertainments considered appropriate only for children and the ignorant. The shadow-play instead called to mind the playful practices of the salon discussed in the previous chapter, and—perhaps especially, in light of the scale of the oratorio—their public counterparts. Toward the end of the eighteenth century, public shadow-plays had found a place among a wider field of scientific spectacles, or philosophical entertainments, that also included automata, electrical demonstrations, and fireworks. The shadow-play was thus an image-projection apparatus associated with higher social circles, more elaborate music, and a learned appreciation for technological wonders.

Shadow-Plays and Fireworks

By the turn of the nineteenth century, entrepreneurial showmen across Europe had taken up the shadow-play as a form of public entertainment. In their advertisements, these showmen styled themselves engineers and artists, and touted both the care and expense that went into their shows and the refinement of their audiences—all strategies crucial to distinguishing the shadow-play from puppets and other lowly forms of entertainment. The "Künstler und Mechanikus" Johann Voack, for instance, advertised "an especially pretty shadow-play fabricated with much hard-work and cost," wherein he spared no expense to make the presentation "according to the taste of the highly-esteemed public."[53] Showmen also often emphasized the musical accompaniment to their shows, which typically comprised songs or other musical performance integrated with the visual display or between the scenes. The illustration by Wilhelm Chodowiecki of a shadow-play in Berlin, discussed in the previous chapter, attests to the elaborate orchestral complement that could be involved, as well as to the well-heeled audience that would attend (figure 3.1). Far from connoting mindlessness, the keyboardist or small orchestra that accompanied a shadow-play helped locate it within a polite culture, and added to its display of skill.

While shadow-play presenters continued to feature comic scenarios or mini-dramas of the sort popularized in the 1770s, these saw growing competition from scenes focused on visually striking landscapes or events. Here, the main interest lay not with the story but with dynamic lighting effects, mechanical contrivance, and the verisimilar imitation of nature. This type of spectacle was pioneered in London by the painter and set designer Philip de Loutherbourg, who introduced the Eidophusikon in 1781 as a novel form of moving picture with music.[54] By the turn of the nineteenth century, however, similar shows had proliferated on the Continent under such names as "Pittoresque theater," "mechanical-optical presentation," and "Chinese shadow-play." Thus, one finds reports such as the following:

> Nothing perhaps gives a more illusory impression than the Chinese shadow-play which was performed by Herr Cavalieri in the theatre of Herr Ritter von Pinetti in Berlin last year [1797] to the great astonishment of the audience. In the area representing the heavens, the spectator could see the dying moon and rising sun represented so naturally that he could not believe his own eyes. In this manner different ani-

> mals and the figures of living and deceased persons, the exhibition of a shower and a hail-storm, are represented very realistically.[55]

Astonishment at the illusion, and disbelief that such realistic representations of nature could be artificially executed, figure centrally in reactions to such shows.

A blend of astonishment and inquiry was characteristic of spectatorship in this milieu, for shadow-plays like Cavalieri's were not only artistic exercises in mimesis: they also belonged to the culture of philosophical entertainments that pressed spectacle into the service of enlightenment. As Barbara Stafford writes, "philosophical entertainments" (the term comes from eighteenth-century England, *physicalische Kunstkabinet* and *magische Experimente* being typical German counterparts) were "exciting ways of doing science by stimulating the eyes," and aimed to "amuse and improve a broad spectrum of European society."[56] Improvement was to come from the demonstration of hidden laws or powers of nature, as well as what such demonstrations intimated about a divine order.[57] Philosophical entertainments thus aimed to be at once enlightening and delightful. They were shadowed, however, by fears of misdirection or fraud, especially by the profit-seeking showman who substituted trickery and self-aggrandizement for truthful investigations of nature.

Joseph Pinetti—owner of the c. 700-seat theater on the Behrenstraße at which Cavalieri performed his shadow-play in Berlin—exemplifies the breed of performer who combined natural philosophy with showmanship and a dash of trickery. Pinetti described himself as a professor of mathematics and natural philosophy patronized by the royal family of France and the court of Prussia (Friedrich Wilhelm II funded the renovations for his theater in Berlin, which was formerly home to Döbbelin's theatrical troupe).[58] But others accused him of being a mere prestidigitator, and indeed while his "chemical art-pieces" and bell-playing automaton Turk were "natural" wonders, he also had recourse to confederates to achieve some of his more explanation-defying feats. A similar tension suffuses the treatise Pinetti offered to the public, *Physical Amusements and Diverting Experiments* (London, 1784). The text dwells mainly on the performance of card tricks, with only scant mention of electricity, chemistry, or other areas of scientific inquiry. Yet, the frontispiece portrays Pinetti as a man of philosophy: he is pictured not performing onstage but as a bust on a pedestal, his lofty gaze suggesting a great mind, cherubs confirming his fêted status, his knowledge

FIGURE 4.5. "Genies place the bust of Professor Pinetti in the Temple of Arts, surrounded by physical and mathematical instruments." Giuseppe Pinetti, *Physical Amusements and Diverting Experiments* (London, 1784), frontispiece. Library of Congress, GV1547.P65.

founded upon the instruments of physics and mathematics strewn below (figure 4.5).

As part of the enlightening project of philosophical entertainments, shadow-plays were frequently paired with another scientific spectacle: fireworks. As Simon Werrett has shown, fireworks had a long tradition of representing royal power in courtly festivals, their explosive display invoking divine miracle and/or battlefield might. The mid-eighteenth century, however, witnessed the emergence of a new genre of "philosophical fireworks" designed to appeal to polite, paying audiences (an appeal that involved minimizing the noise, smell, and filth of the fireworks and of the audience typical of festive outdoor displays).[59] Programmed together, shadow-play and fireworks provided complementary demonstrations of nature's powers, allowing the philosopher-showman to theatricalize his control of light and fire. In 1791, for example, Herr Pleiner exhibited a "great German shadow-machine" and "mechanical fire-work decorations." The former included comic scenes and the visually striking effects of dancers and quick transformations; the latter comprised tributes to Prussian and Austrian rulers. But Pleiner's advertisement prominently promised "in shadows" and "in fire"—the objects to be represented appearing in smaller font below the elements of nature to be controlled (figure 4.6).[60]

In Germany, shadow-plays and fireworks came into their own as forms of scientific spectacle for paying audiences around 1790. Interest in the type of performance as a domestic, amateur activity was already burgeoning in 1770s Berlin, however, when Zelter—who spent his life in the city—was in his formative teens and early twenties. In his autobiography, Zelter recalled attending the almost daily "concerts with beer and tobacco" performed at the public gardens by military and city musicians.[61] Through these events Zelter got to know the multi-instrumentalist Lorenz George, for whom he worked as he furtively pursued musical training alongside the more stable, reputable career of master mason inherited from his father.[62] Zelter was especially taken with George's home, located in a section of Berlin where a house and garden were affordable on modest means. Zelter recalled several great parlors with "walls bedecked with the usual musical instruments," one of which additionally held an assortment of nonmusical instruments: numerous woodworking tools, fireworks equipment, an electricity machine, and "a hundred things that one rarely sees together."[63] In this room we glimpse a private counterpart to what would become familiar as the philosopher-showman's *Kunstkabinet*.

George put his fireworks equipment to use in the summers, when they formed part of the convivial activities (along with various athletics and improvised comedies) that brought friends together in his private garden. As Zelter recalled:

> Since George was also concerned with the nature of fireworks, wherefore each of his friends had to help acquire paper, gunpowder and the like, there were now and then fireworks at whose lighting kettledrums

were sounded and struck. He called this his eye-concert and, like a child, long looked forward to the moment when he would see all these things light up and play in the air. Paper kites of uncommon size were fashioned, illuminated with artificial fire and in the autumn, sent over the stubble field into the air trailing a long tail of fire. The jubilation there when everything ran well was gratifying, and many days afterwards the subject of conversation.[64]

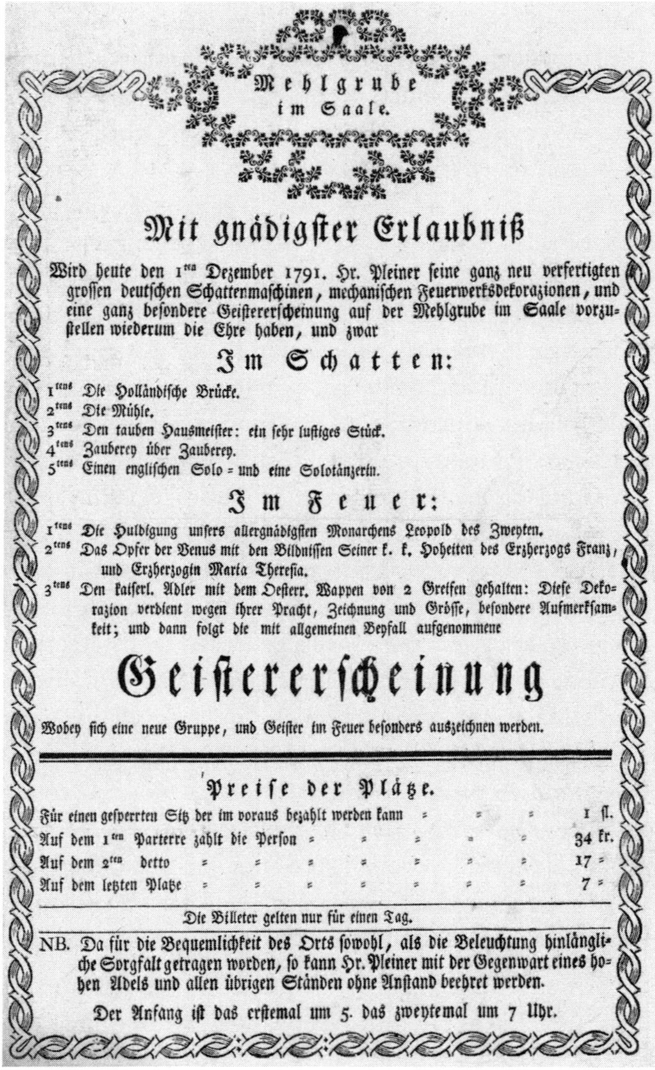

FIGURE 4.6. Poster advertising Herr Pleiner's new shadow-play, fireworks and ghost show, Vienna, 1791. Wienbibliothek im Rathaus, Plakatsammlung, D-64522.

In pyrotechnics, George found enjoyment "like a child" but also a means to engage with "the nature" of fireworks. While representing objects might have made the fireworks especially impressive, the main concern was with successful control over the physical materials: "everything running well" was the occasion for jubilation and further conversation. The play of sensory impressions was thus both pleasurable in and of itself, and a manifestation of the natural forces and technological mastery that made it possible. As a temporal art of fleeting visual impressions, moreover, fireworks were analogous to music—a concert for the eyes.

Zelter described a similar mix of childish wonder and higher intellectual engagement in musical performance. Recalling his first operatic experience—attending Graun's *Phaeton* at age eleven or twelve—he described the orchestra as a technological marvel:

> I saw the tones of the singer coming, as it were; but the orchestra on the whole was a huge, pleasurable mystery to me. I was among the musicians, each of whom played *one* instrument, and yet I heard not *one* but rather the orchestra itself, which I imagined as an enchanting sound box, as a type of organ, ringing and resounding against me as *one* whole. By this observation my gaze became lured to the stage, and I swam in a sea of joy.[65]

The orchestra defied understanding: it appeared as many yet sounded as one. Though initially astonished by this phenomenon, which he could not explain but only experience, Zelter progressed from wonder to "observation": the orchestral wonder led him to the stage—to the drama from which, according to standard theories of the period, he might learn such virtues as sensibility and moral uprightness.

Zelter's autobiography illustrates the possibilities for productive exchange between musical listening and spectatorship of philosophical entertainments. Musical and scientific performances both were (or could be) highly sensory affairs that produced a fascination with their mechanisms—mechanisms that, at least initially, exceeded one's understanding. What was represented, meanwhile, mattered comparatively little. The purpose of representation had less to do with storytelling or arousing emotions than with demonstrating technological mastery and occasioning wonder. What was imitated was secondary to the fact of imitation, and to the question astonishing similitude prompted: how was this accomplished?

Haydn's *Creation* as Philosophical Entertainment

Haydn's *Creation* posed an aesthetic challenge. Noting that Haydn "playfully calls forth wind and weather, snow and hailstones," Zelter explained, "one is accustomed to see the passionate upheavals of an actual person when one thinks of such concepts; or a soul in turmoil; but these are matters that should not be sought here."[66] In a review of Haydn's later (similarly representational) oratorio *The Seasons*, Zelter addressed the same problematic expectation, which continued to limit critical appreciation: "one could be quickly enough finished with a judgment of this important work if one were to subject it to that theory according to which music is supposed to be nothing else but the language of feelings when it wishes to communicate them."[67] Rather than criticize Haydn for failing to adhere to existing aesthetic theory, Zelter repeatedly emphasized the need for a totally new theory by which to judge his oratorios.

Early on, Zelter sought such a theory in the realm of science. In his first review of *The Creation*, published three months before Triest's comments, he portrayed the oratorio as a natural history collection and educational treatise: "the physical side of nature, with its expressions, so far as they concern tone and movement, is collected here as thoroughly as for an encyclopedia [*Lexicon*]; and he who does not want to go himself, needs only look here, and if he understands how to look, he will find."[68] The metaphor located value in Haydn's exhaustive collection of musical representations of nature, in the accuracy of those representations, and ultimately in the didactic significance of the achievement: one could now learn about the physical side of nature by studying Haydn's oratorio.

In his subsequent articles on Haydn's oratorios, Zelter gradually shifted emphasis from the natural phenomena Haydn represented to the musical medium. The shadow-play provided a conceptual pivot point in this shift. Zelter's description of *The Creation* "parading before the inner eye, like a fine shadow-play," suggested experiencing the oratorio literally as a series of moving images of nature, with music being the means of animating these images in the mind ("movement and repose are made to come alive . . . by the art of the music"). In his later review of *The Seasons*, however, Zelter moved beyond the visuality of the listening experience. While finding "the total impression of *The Seasons* [to be] like a gallery, a suite of paintings," he also likened the work to dynamic visual spectacles made from fugitive materials. It was in connection with these dynamic visual media that he identified the true aim of Haydn's oratorio: "the whole intention is therefore sensuous, is supposed to act sensuously and to disappear from the senses even as it appeared; it is a higher play for the ear and may be compared to a shadow-play and with fireworks . . ."[69]

Rather than suggesting merely trivial spectacle, Zelter's shadow-play and fireworks invoked the combination of sensory play and intellectual work characteristic of philosophical entertainments. What shadow-plays and fireworks did by addressing the eyes, *The Creation* did by addressing the ears. Likening *The Creation* to philosophical entertainments thus suggested conceiving music—like light and fire—as one of nature's indwelling powers: as fugitive, mysterious, yet available to be mastered through the application of intellect and technology. Such a conception of music offered an alternative to the "language of emotions," which figured music as a means of expressive communication between people. Instead, it figured music as an auditory spectacle at which to marvel—not merely for pleasure but also to gain knowledge.

The idea that musical composition and performance constitute an artful play with nature's powers (and more specifically that Haydn achieved such play in his oratorio) draws support from the depiction of chaos with which *The Creation* opens. As scholars have argued from a number of perspectives, Haydn's musical chaos presents the materials of music as raw elements: the instrumental sonorities, melody, and harmony behave fugitively and erratically until finally brought to order in the form of a tutti C-major chord, at the moment of the creation of light.[70] Zelter heard Haydn's chaos in precisely such terms of raw matter and power: "a gigantic *unisono* of all instruments, at the same time a light-less and formless mass, are suggested to our imagination. From it single notes come forth, which in turn spawn others . . . Huge forces grate against each other and begin to gestate, and occasionally, as if fortuitously, they dissolve harmonically and then sink back into darkness."[71] For Zelter, Haydn's representation of chaos also triggered the disbelief that accompanied seemingly impossible human-technological accomplishment—in this case, the simulation of disorder by means of harmony, melody, and rhythm: "despite the appearance of impossibility and contradiction, the marvel is in fact the most poetic and hence the best part of the whole plan."[72] Rather than a form of illusion suitable only for the ignorant, the marvel of Haydn's chaos became the epitome of his creative genius.

Marvelous achievement could also be located in Haydn's tone-paintings. Here, too, Zelter found evidence of Haydn's genius. "Only a vivid imagination," he wrote, "could create picture after picture, describing natural occurrences as easily and with a few strokes as they were easily created in the first place."[73] Other listeners, however, articulated amazement at Haydn's tone-paintings even more clearly. After attending the first public performance of *The Creation* in Vienna, Joseph Richter (the same who authored the libretto *Das schöne Milchmädchen, oder Der Guckkasten*) expressed with enthusiasm his wonder at Haydn's exact imitations of nature:

> for the life of me I wouldn't have believed that human lungs and sheep gut and calf's skin could create such miracles. The music all by itself described thunder and lightning and then you heard the rain falling and the water rushing and the birds really singing and the lion roaring and you could even hear the worms crawling along the ground.[74]

By placing tone-paintings before their verbal elucidation, Haydn enabled listeners to judge the intelligibility of "music all by itself." But where Triest's magic lantern suggested the result was utter failure, Richter marveled at the resounding success of Haydn's auditory simulations, which gained significance from being so improbably accomplished with human lungs, sheep gut, and calf's skin. Haydn's early nineteenth-century chronicler Giuseppe Carpani echoed Richter's amazement in a verse honoring the composer: "But that a man the work of God, stupendous, rare,/Should try to equal with pictorial notes,/And render clear and present to the aware,/This seemed impossible to human mind./This, Haydn, thou hast done . . ."[75] Here, Haydn's success with "pictorial notes" puts his creative power on par with God's.

As metaphors for Haydn's oratorios, then, shadow-plays and fireworks pointed to a type of experience shared by many enthusiasts of the oratorio, particularly those who embraced its tone-painting: the mixture of uncertainty, awe, and curiosity felt on confronting an inexplicable or seemingly impossible human-technological achievement. Such experiences could lead to a better understanding of nature and appreciation for its divine order. They could also inspire hypotheses of supernatural intervention. The reception history of *The Creation* finds the latter often in evidence. At a performance in 1808—Haydn's last public appearance—Haydn gestured toward the heavens during the applause that followed his musical creation of light. Taken to mean "it comes from up there," the famous gesture reinforced the conflations of musical with divine creation already in circulation, and suggested that higher powers had a hand in the "miracle" of his musical work.[76]

The framework of philosophical entertainment provided a way to appreciate vivid imitations of nature produced skillfully and miraculously from musical instruments and tones; but it also suggested the dangers of such imitations, which instead of leading to knowledge could lead to superstitious belief. The image of the divinely inspired artist, in fact, stood at odds with the image of the natural philosopher. For rather than being celebrated, philosopher-showmen who flirted with the supernatural became the focus of campaigns against superstition and charlatanism. With the rise of fraudulent ghost-raisers in the 1770s and 1780s (a subject of the next chapter), efforts to disabuse the public of belief in ghosts gained momentum, and a chief means for exposing false conjurors was to explain the mechanics of image projection. While the magic lantern was the instrument of choice for ghost-raisers and a primary target of demystifiers, shadow-plays too were discussed and tarnished by association with dishonest conjurors. Johann Heinrich Helmuth, for example, added the shadow-play to the fourth edition of his anti-superstition treatise *Volksnaturlehre zu Dämpfung des Aberglaubens* [People's natural lesson to curb superstition], cautioning that "if an ignorant person sees such ideas: he can easily imagine that they were produced by the arts of magic."[77] Coming on the heels of a section on the use of the magic lantern in ghost-raisings, the passage goes on to expose the techniques by which Cavalieri achieved the effects in his shadow-play. Although Helmuth stops short of accusing Cavalieri of charlatanism, he nonetheless treats the shadow-play as an entertainment that preys on the ignorant: audience members who read his treatise, he implies, will cease to be amazed by the spectacle and will be better off than their ignorant counterparts. As concerns about their ability to inspire false beliefs overshadowed the enlightening work of technological wonders, their promise as models for a new music-aesthetic theory likewise faded.

Nocturnorama: An Epilogue

The debate over the aesthetic worth of Haydn's *Creation* was simultaneously a debate over the future of music. As Matthew Head has demonstrated, Triest's and Zelter's comments on the oratorio participated in a wide-scale "centennial enterprise"—an effort to distinguish the eighteenth century from the new century to come.[78] For Triest, Haydn's symphonies represented the crowning musical achievement of the 1700s, and *The Creation* an aberration that—despite its turn-of-the-century popularity—should not be counted toward the composer's legacy. For Zelter, *The Creation* extended Haydn's innovations in instrumental music while also being a wholly new form, making it a promising progenitor of new directions for the future.

These critics' takes on *The Creation* also contested the status of moving images and the future of audiovisuality. Triest's ideal of "free" listening meant that any externally fixed images automatically were less than, and lessened, musical experience. This was not to say that listeners should focus strictly on the sound and structure of "pure instrumental music," but rather to insist on the value of inner sensations. As he argued, the "greatest effect" was "when the listener—even without auxiliary elements such as ballet, marches, etc.—believes that he perceives, in the composition that is being performed, the expression of specific ideas or feelings that have simply not been expressed, and easily supplements the text, as it were, with his thoughts."[79] Zelter, by contrast, set spectacles for eye and ear on equal footing, as exemplified by his interchangeable notions of fireworks as an eye-concert and music as fireworks. He clarified the theoretical basis for this sensory equality in his autobiography, recalling an occasion in 1783 when he argued that "there is only one Art; painting and music are only different fields, part of this general Art; one must know the boundaries, but also how it looks from the other side."[80] Zelter's turn-of-the-century writings on Haydn's oratorios extended the "one Art" of painting and music to moving images, suggesting that the relations among these fields should be further explored in the new century.

As if to answer the call, a certain Porro family announced in 1836 "a new sort of concert [to] entertain the fashionable society of Paris this winter. All that the music expresses, during these concerts, will be rendered visible through painted transparencies of superior quality. Haydn's *Creation* is in rehearsal and, accompanied by the appropriate phantasmagorias, will no doubt doubly captivate the sense of the audience."[81] The Porros linked their "new sort of concert" to a proliferation of visual entertainments. In the wake of the panorama, introduced in late eighteenth-century London and from there all over Europe, a wave of other "-oramas" appeared. Each distinguished itself by slightly different subject matter and technical means, but all aimed for a striking illusion of nature. As the Porros observed in their announcement, the public already had the Panorama, Diorama, Cosmorama, Neorama, and Uranorama. "What was missing was the Nocturnorama," as they called their union of music with painted transparencies, promoting it at once as an innovation and the ultimate form of "-orama" entertainment.[82]

The selection of Haydn's *Creation* for this experiment in illusionistic spectacle attests not only to its enduring popularity but also to its ongoing identification with moving-image apparatuses—its tendency to stimulate listeners' imaginations along technological lines.[83] The reaction to the experiment likewise shows the evolution of music-image relations. The German author August Lewald expressed reservations about putting illuminated images—which were "more suited to gay and sentimental diversions"—to "this great work."[84] Like Triest, Lewald regarded moving images as a lowly form of entertainment, to which "aesthetics" could have nothing to say.

But rather than regarding *The Creation* as itself already on the level of moving-image entertainments, as did Triest, Lewald held it to be a great musical work that produced a "magical effect" on sensitive listeners equal to that of Haydn's symphonies.[85] In this regard, Lewald was in agreement with Zelter. Citing the comparison of *The Creation* to a magic lantern, Lawrence Kramer has credited Triest with having "virtually the last word" on the subject of musical representation, his "proto-formalist" attitude heralding the belief in purely musical logic that came to "set the terms for serious thinking about music."[86] But Lewald's comments—his separation of trifling moving images from Haydn's great oratorio—suggest that Triest's was far from the last word. Rather, even while the particulars of Zelter's metaphoric network lost currency, it remained possible for serious listeners to appreciate the oratorio, tone-paintings and all. That Haydn's music could conjure the first days of the world all on its own was a kind of magic, and crucial to cementing the status of *The Creation* as a musical work both dubious and great.

FIVE

Beethoven's Phantasmagoria

BEGINNING WITH THE QUIET THUDS OF THE TIMPANI AND ENDing with a fortissimo C-major chord in the full orchestra, the transition into the finale of Beethoven's Fifth Symphony dramatizes its trajectory from minor to major, struggle to victory, darkness to light, and helps turn the multi-movement work into a heroic journey and unified whole. At least, these are the functions of the passage that have been emphasized by a long interpretive tradition. Yet, this and other passages of the symphony carry not only such narrative and formal implications, but also spatial ones. Around the turn of the nineteenth century, illusions of motion through space—both acoustic and optical—became central to transporting spectators into other worlds. These illusions produced experiences of immersion strikingly different from the distanced spectatorship of peepshows, magic lanterns, and shadow-plays, with a correspondingly updated power to induce belief. Deployed in opera from the mid-eighteenth century, such manipulations of spatial perception furnished the basis for a form of entertainment that emerged in the 1790s and remained popular well into the 1800s: the phantasmagoria.

Since its coinage for ghostly entertainments, the term *phantasmagoria* has acquired a rich set of metaphorical and critical meanings. Often it serves to conjure spectral images changing quickly or possessing a dreamy illogic, whether these images be in the world or in the mind. Indeed, as Terry Castle has argued, the term underwent a process of internalization over the course of the nineteenth century, helped along by rationalist accounts of ghosts as products of human imagination rather than of supernatural agencies, and the confusion of inner and outer realms remains crucial to its evocative resonances.[1] *Phantasmagoria* also has a specifically musical application thanks to Theodor Adorno's writings on Wagner. Borrowing the term from Karl Marx, for whom it described the concealment of labor in the form of the commodity, Adorno applied it to musical passages that efface the labor of composer and orchestra: the "phantasmagoria par excellence" was thus "the image of loudness from afar," an acoustic illusion achieved by scoring for high woodwinds and no bass instruments, among other techniques.[2]

Lost or attenuated in these usages of *phantasmagoria* are the sensory effects and philosophical concerns that characterized the spectacle in early nineteenth-century Germany. Here, the "phantasmagoria par excellence" was not a static illusion of distance but rather a dynamic process of gradual approach. The looming effect extended an arms race between audiovisual techniques for inducing belief in spirit appearances and discursively mounted defenses against such belief; it entered an ongoing debate about spirit-seers—as those who believed spirits could be perceived were called—in which philosophers sought to explain the phenomenon of mystic visionaries like Emanuel Swedenborg, or fraudulent ghost-raisers such as Schröpfer and Cagliostro. A source for the ghostly imagery E. T. A. Hoffmann employed in his famous review of Beethoven's Fifth Symphony, these techniques and debates also shed light on the implications of his claim that Beethoven's instrumental music turns us into "spirit-seers," and the concerns animating his novel analytical search for hidden unity.

Robin Wallace has observed that the symbolic freight Beethoven's persona and music now carry "makes it difficult to *listen* to him the way that Hoffmann and his contemporaries did."[3] Our distance from these listeners has to do not only with the "heroic" lens through which we tend to approach Beethoven's music, however, but also with the different "perceptual world" (to use Benjamin's term) in which we listen. It has become common to regard nineteenth-century technological innovations as fundamentally transforming the senses; according to Jonathan Crary, the 1830s saw the meanings and effects of images newly conditioned by the "overloaded and plural sensory environment" of proliferating optical technologies.[4] Music offers a more sensitive barometer of changes to perceptual practice. As the preceding chapters have shown, it was from the 1770s onward that moving-image technologies became increasingly salient in the plural sensory environment of musical performance and reception. This chapter recovers another crucial piece of the perceptual world of Beethoven's early listeners to develop a hearing of the Fifth as phantasmagoria. In so doing, it resituates the work at the intersection of material culture and philosophical discourse—of popular entertainment, fine art, and scientific practice—and it locates the conditions of its aesthetic force in the spatial manipulations and metaphysical uncertainties of the early nineteenth century.

Spirit-Seers

In the mid-eighteenth century, word of Emanuel Swedenborg—a man who reported visions of the spirit world and seemed to possess telepathic powers—prompted excitement throughout Europe. Immanuel Kant learned of the visionary in the early 1760s, and became intrigued by the promise of empirical evidence for metaphysical realms. He investigated the matter, recruiting an acquaintance to make inquiries in Swedenborg's hometown of Stockholm and reading the seer's eight-volume work *Arcana Coelestia* (1749–1756). The result was the damning treatise *Träume eines Geistersehers* [*Dreams of a Spirit-Seer*, 1767], in which Kant not only branded Swedenborg a *Schwärmer*—a derisive term for self-deluded religious enthusiasts—but also indicted the entire field of metaphysics as then practiced. In Swedenborg's meticulously detailed and seemingly rational accounts of the spirit world, Kant recognized the type of systematic investigation of the immaterial world in which reputable philosophers also engaged. The spirit world, Kant concluded, could not be known because "there is no data for such thought";[5] at best, influences from the spirit world could be "transformed into shadow-pictures [Schatten-Bilder] of sensible things," at which point there was no way to distinguish them from "wild figments of the imagination" [wilde Hirngespinste].[6] Were that not enough, Kant added that if madness were not necessarily the cause of spirit-seeing, it would certainly be the result of such supernatural communication should it occur. Henceforth, Kant concluded, metaphysics should cease to make claims about the spirit world and instead become "a science of the *limits of human reason*."[7]

When Kant published *Dreams of a Spirit-Seer*, the term *spirit-seer* was reserved for people like Swedenborg who claimed to communicate with spirits in the privacy of their own visions. By the 1790s, however, the term was also applied to people who produced spirits for public view by means of technological trickery. Thus, the central episode of Friedrich Schiller's novel *Der Geisterseher* [*The Spirit-Seer*, 1787] concerns a ghost-raising ceremony wherein an apparition is manufactured by magic lantern. The continuity between spiritual visions and optical presentations—already hinted at by Kant's characterization of spirit perceptions as "shadow-pictures" indistinguishable from wild imaginings—became a distinctive feature of discourses on spirit-seeing. In an essay entitled "Emanuel Swedenborg: Greatest Spirit-Seer of the Eighteenth Century," Herder concluded that the mystic did not receive divine knowledge but rather "saw and believed with conviction these fantasies which sprang from his inmost being, this conviction based upon inner appearances represented as present to the senses. Heaven and hell were from and in him, a *laterna magica* of his own thoughts."[8] Even those who maintained the reality of spiritual visions suggested they resembled magic-lantern images and figments of the imagination. According to the mystic philosopher Karl von Eckartshausen, there were three kinds of spirit appearance or apparition (*Geistererscheinung*):

> The first one is purely artificial, consisting of an optical deception. The second kind is produced through the images of the imagination, that is, by the imagination creating an external image outside of the body. And the third is the true spirit apparition, visible only to the inner sense, and transformed by this very inner sense into an image for the outer senses, which is in fact the true apparition.[9]

While all spirit appearances manifested outside the body (in contrast to mere mental images), one was produced by light hitting the eye, one by the errant imagination, and one by the "inner sense" responding to true spirit influences. The first, "purely artificial" spirit appearance, Eckartshausen proceeded to detail, was made with the magic lantern.

The newfound place of magic lanterns in discussions of spirit-seeing owed to the performance of ghost-raising ceremonies, which came to public attention around 1770. In Germany, Johann Schröpfer rivaled the great magus Cagliostro for fame and infamy as a necromancer. Claiming possession of the secret knowledge of Freemasonry, Schröpfer began performing ghost-raisings in 1768 in a coffeehouse in Leipzig. He continued until his death in 1774, bringing his magic to cities including Berlin, Frankfurt am Main, and Dresden. In Dresden he was even invited to raise a ghost at court, and reportedly succeeded in convincing those assembled of his powers. The event took place in 1773, and prompted the English visitor N. W. Wraxall to observe the metaphysical inclinations of his German counterparts, who were "more open to superstitious impressions, than with us. The Germans, almost universally, even those of the soundest and most cultivated understandings, believe in the existence of familiar spirits; in whose train follow witches, ghosts, and the whole family of invisible agents."[10]

Schröpfer and his ilk prompted German intellectuals to scrutinize the techniques of ghost-raising, to counter the "almost universal" belief Wraxall observed. In "natural magic" treatises—the variety of popular scientific literature centered on explicating phenomena that "seem to surpass the natural powers of the body," as we saw in the previous chapter—philosophers taught that ghost-raisers used a magic lantern concealed in a box, its projected image reflected by means of a mirror onto a cloud of smoke.[11] Unlike a traditional magic-lantern slide, the slide used by a ghost-raiser featured figures surrounded by black, so that the figure alone was projected without any surrounding light. The resulting ghostly appearance can be seen on the title page of Johann Samuel Halle's natural-magic treatise, *Magie, oder die Zauberkräfte der Natur*, where it appears in the context of a ghost-raising ceremony (figure 5.1).

The projection of ghostly images and concealment of the apparatus were not the only requirements for a convincing ghost-raising, however: sounds were equally important. Poorly executed sounds could ruin the optical illusion. Thus, a witness quoted in Christlieb Benedict Funk's natural-magic treatise reported that he had almost succumbed to belief when the ghost began to speak: this quickly cured him of his shock and fright, for "I could perceive quite clearly that the spirit answered in Yiddish with German words interspersed; the heavenly spirits surely speak pure Hebrew."[12] Well-executed sounds, on the other hand, not only preserved the optical illusion but also made audiences more susceptible to belief. Halle—describing the arrangement pictured in figure 5.1—explained that the ghost emitted plaintive howls and answered questions by means of a tin-plated tube, through which an assistant spoke from an adjacent room. According to Halle, this transmitted voice worked to defeat skepticism: "we hear his hoarse, unnatural voice and immediately our philosophy disappears, and we tremble in the [magic] circle rather than laughing at the Magus's wrinkled brow."[13]

Storm sounds too were integral to ghost-raisings, with thunder being considered an especially powerful auditory tool for frightening audiences into belief. To produce thunder, ghost-raisers typically used a large, tightly stretched drumskin, beaten either with sticks or bare fists. As Johann Peter Eberhard observed in his *Abhandlungen vom physikalischen Aberglauben und der Magie* (1778), such noises had a powerful psychological effect: "there occurs preliminarily a rumbling, and now the audience is more dead than alive, their imagination now highly agitated and ready to accept any impressions the magician wants to give it."[14] By thus "put[ting] the audience in a state of awe and terror" before the ghost's appearance, ghost-raisers rendered viewers "incapable of noticing the fraud."[15]

Ghost-raisers, then, were experts in technological and psychological trickery. By explaining the natural causes and psychological effects of the ghost-raisers' sights and sounds, the authors of natural-magic treatises sought to inoculate readers against his tricks—to equip reason to triumph

Magie,
oder, die
Zauberkräfte der Natur,
ſo auf den Nutzen und die Beluſtigung
angewandt worden,
von
Johann Samuel Halle,
Profeſſoren des Königlich-Preußiſchen Corps des Cadets
in Berlin.

Mit 9 Kufertafeln.

Erſter Theil.
Zweyte Auflage.

Berlin, 1784.
bey Joachim Pauli, Buchhändler.

FIGURE 5.1. Johann Samuel Halle, *Magie oder die Zauberkräfte der Natur*, vol. 1 (Wien: Johann Thomas Edlen von Trattnern, 1787), title page. New York Public Library, 3-OAL.

over the senses, emotion, and imagination. They taught a skeptical and investigative attitude toward the miraculous: rather than believe one's eyes, one should assume and search out hidden natural causes. Christoph Friedrich Nicolai—an important publisher of natural-magic treatises—succinctly captured the value of such an attitude in a report on his own visionary experiences in 1791. Had he been superstitious, he would have been driven mad with fear; had he been attached to the marvelous, he would have aggrandized himself as a spirit-seer; instead, he calmly and rationally traced his visions to their natural cause in disease. "The advantage of sound philosophy and deliberate observation may be seen," he concluded, "both prevented me from becoming either a lunatic or an enthusiast."[16]

The skeptical, investigative attitude cultivated toward ghosts likewise found practical application at a ghost-raising ceremony in Berlin on March 30, 1789. The ghost-raiser was Phylidor, a conjuror who publicly gave magic shows "imitating the so-called black arts" with mechanical and magnetic contrivances that appeared to respond to verbal command. By private arrangement, however, Phylidor offered to use his magic to raise the dead. Phylidor's ghost-raisings closely resembled those described in natural-magic treatises. First, he frightened spectators with a preliminary rumble. Then he called up two ghosts readily identified from their widely disseminated likenesses: Voltaire, the French *philosophe* who had died a decade earlier; and Frederick the Great (Friedrich II of Prussia), the enlightened monarch only three years in the grave. A contemporary illustration of Phylidor conjuring Friedrich II captures the ambiance of smoke and darkness (figure 5.2).

The *Berliner Monatsschrift*—a vehicle of Enlightenment rationalism—published two eyewitness accounts of Phylidor's March 30 ghost-raising; each witness saw through the conjuror's tricks, thanks to his skepticism and careful observation. As one, von der Recke, reported, he approached the event "with the firm decision to uncover the fraud wherever possible, however deeply it lay hidden," and concluded by calling upon everyone, when presented with such supernatural appearances,

> to examine and investigate them, and get to the root of the matter, in order to convince themselves that in all such performances, if the same unnatural things appear, they still have the most natural causes, even if at times this escapes our notice.[17]

The journal editor added a note wishing that everywhere in Germany should have "a society of such unprejudiced and incredulous spectators and of such insightful and truthful researchers."[18]

Phylidor's reception in 1789 Berlin marked a turning point in the history of spirit-seeing: a culture of skepticism and investigation now imperiled anyone banking on credulous spectators. Informed of Phylidor's

FIGURE 5.2. Phylidor conjuring Friedrich II in Berlin, 1789, by Daniel Chodowiecki. Reproduced from Walter Stengel, *Guckkasten: Altberliner Curiosa* (Berlin: De Gruyter 1962), 176.

ghost-raising, the state police determined that it "offended against religion and good morals," forbade his performances, and ordered him to leave the city.[19] But Phylidor's Berlin reception also suggested a different kind of performance in which ghost-raising would be allowed, for his ceremony was seen to have a certain educational value: such "dabbling in Swedenborg's, Schröpfer's and Cagliostro's craft" was useful, the *Berliner Monatsschrift* reported, "partly in that it reveals the variety and mechanism of the fraud; partly in that, from the conduct of the audience, the disposition and receptivity of the vulgar for such follies can be judged."[20] Nine months after being expelled from Berlin, Phylidor reemerged in Vienna offering a new kind of show: he now advertised himself as a physicist-philosopher exposing fraudulent ghost-raisers by replicating their ceremonies. The result was a type of ghost show that came to be called *phantasmagoria*.

Phantasmagoria

The phantasmagoria thus repackaged and repurposed ghost-raisings as public entertainments and debunking exercises in the service of enlightenment. Phylidor called his new show "Phantasmorasi, or natural Spirit Appearances" (Phantasmorasi, oder natürlicher Geister Erscheinungen), as well as "Spirit-Appearances in the Manner of Schröpfer and Cagliostro" (Schröpfersichen und Cagliostroischen Geister-Erscheinungen) after the two famous necromancers (figure 5.3).[21] The Parisian scientist and showman Étienne-Gaspard Robertson, who copied Phylidor's show and coined the term *phantasmagoria*, similarly connected his performances to the history of superstitious belief, advertising "apparitions of Spectres, Phantoms and Ghosts, such as they must have appeared or could appear in any time, in any place and among any people."[22]

Even as these showmen presented their performances as exposés of the tricks behind the ghost-raisings of old, however, they also developed new techniques. The drive to technical innovation stemmed in part from the scientific prestige new inventions carried and the commercial appeal of novelty. Thus, one advertisement announced: "long occupied with bringing the artificial ghost-appearance to more perfection, Mr. Phylidor has after many experiments found an entirely new method . . ."[23] While Robertson claimed his ghosts were the same as those of any time or place, he simultaneously filed a patent for his new and improved magic-lantern apparatus for projecting ghosts, which he called the Fantascope. Both showmen preceded their ghost presentations with other "new experiments" in natural magic, using the phantasmagoria as a grand finale to their demonstrations of cutting-edge scientific knowledge.

Most importantly, there was a need for new techniques to make even skeptical and inquiring spectators believe they were seeing ghosts. Such belief was central to both the pleasure and the educational mission of the phantasmagoria: its showmen were in the business of debunking supersti-

Nachricht.

Heute und alltäglich um 7 Uhr werden öffentliche Vorstellungen der neuesten Experimente aus der natürlichen Magie; nebst die mit Beyfall aufgenommenen

Schröpferischen, und Cagliostroischen

Geister=Erscheinungen

im Phylidorschen, Physikalischen Kabinet gegeben werden.

Herr Phylidor hat die Ehre hiedurch das hochschäzbareste Publikum zu benachrichtigen, daß er nicht nur allein seinen bisher bewirkten Darstellungen mehre Vollkommenheit beygebracht hat, sondern auch sich beschäftiget dem ganzen die schon längst gewünschte Vollkommenheit zu geben, um die schäzbaren Zuschauer in der angenehmsten Täuschung zu erhalten, ohne des Lichts beraubt seyn zu dürfen, zu welchem Endzweck er bereits sein Kabinet einrichtet, und mit nächstem die Ehre haben wird, die bisher nur im Finstern gesehene Geister, in den herrlichsten Elisäischen Feldern zu produzieren.

Denen verschiedenen fremden Herrschaften und Kunstfreunden, welche täglich um die Beschreibung meiner jezigen Vorstellungen zu mir schicken, eine Idee zu geben, wird aus dem Ganzen ein kleiner Auszug gemacht, weil aus Mangel des Raums nicht wohl möglich ist, eine ausführliche Beschreibung von allen vorzustellenden Experimenten beyzufügen.

Die Zuschauer versammeln sich in einem schwarz behangenen Saal, der Operationsplatz stellt einen Tempel vor, sobald die Operation vor sich geht, so höret man von ferne den Donner sich näheren, der sich nach und nach immer mehr verstärkt, und mit Hagel, und brausendem Wind begleitet wird, es erlöscht ein Licht nach dem andern von sich selbst, und es steigen aus den Lampen feurige Flammen empor, hierauf erlöschen die Lampen, und man ist sodann in einer undurchdringlichen Finsterniß, verschiedene Geister erblickt nun das Auge des Zuschauers von aller Gestalt und Größe, sie flattern in einem Zirkel herum, und verschwinden wieder, in jeder Vorstellung erscheint die Aehnlichkeit verschiedener der Gesellschaft sehr bekannter Personen in der Gestalt des so allgemein angenommenen Begriffes eines Geistes, ein jeder von Ihnen entsteht, und verschwindet auf eine andere Art, einer kommt aus dem Fußboden empor; der Zweyte stehet plözlich auf einmal da, der dritte bildet sich aus einer grauen Wolke, und wird nach und nach so körperlich, daß man deutlich die vorgebliche Person in der Gestalt erkennt. Eine entsteht in der Entfernung, und so täuschend, als wenn selbe 200 Schritt weit entfernt wäre, und nähert sich bis zum greifen, jede Erscheinung macht einige Schritt vorwärts gegen die Zuschauer und verschwindet.

Alle diese sogenannten Geistererscheinungen machen weder auf den Geschmack, noch auf die Geruchsnerven der unschätzbaren Zuschauer einen nachtheiligen Eindruck.

Außer dieser so seltenen und angenehmen Unterhaltung, wird man die Ehre haben, ehe die Erscheinungsoperation angehet, die ersten mechanisch = magnetischen Experimente zu zeigen, durch welche man dem resp. Auditorium nacheinander folgende angenehme Ueberraschungen zu verschaffen suchen wird.

Eintrittspreise.

Im ersten Rang.	40 kr.
Im Zweyten.	20 kr.
Für einen abgesonderten Platz aber, der besonders voraus bestellt und gesperrt werden kann, zahlt die Person.	1 fl. —

Der Schauplatz oder das Kabinet ist im kleinen Loprestischen Hause am Kärntnerthor über dem Kaffeehause Nro. 1190. von der Bastey hinein, allwo auch die gesperrten Plätze und Eintrittsbillets für jede Vorstellung abzuhohlen und die Privatvorstellungen täglich zu machen sind.

Der Anfang ist präcise um 7 Uhr, und dauert 1 und eine halbe Stunde.

FIGURE 5.3. "Today and everyday at 7 o'clock/Presentations of the newest experiments from natural magic/and the much applauded/Schröpfer-style and Cagliostro-style/Spirit-Appearances/in Phylidor's physical cabinet." Broadside, Vienna, 1790 or 1791. Wienbibliothek im Rathaus, Plakatsammlung, D-64522.

tious beliefs by making avowedly false ghosts feel real. This called for more sophisticated stratagems than those employed by the previous generation of ghost-raisers. The phantasmagoria showman's tricks were thus calibrated to the perceptual competencies of turn-of-the-nineteenth-century audiences, and designed to defeat their capacity for rational explanation.

The resulting innovations centered on illusions of motion through space. Such movement began with sound: in place of a single frightening rumble, Phylidor opened his shows with "distant thunder coming closer, becoming gradually stronger and stronger, accompanied by hail and roaring wind."[24] As Phylidor further described in advertisements, an invisible agent seemed to move about the hall, as "one light after another extinguishes itself and fiery flames rise up from the lamps; then the lamps go out and one is in impenetrable darkness . . ."[25] One then saw "several spirits of all shapes and sizes, they flutter around in a circle and disappear again . . ."[26] Finally, Phylidor presented figures like the one in his Berlin ghost-raising, newly mobilized:

> The likeness of various people well-known to society appears in every presentation, in the form of the concept of a spirit so universally adopted. Each spirit arises and disappears in a different way: one comes out of the floor; the second is suddenly there all at once; the third forms from a gray cloud, and little by little becomes so corporeal that one clearly identifies the alleged person in the shape. One emerges in the distance, and so deceptively, as if it were 200 strides away and approaches in order to attack; each apparition strides forward toward the spectators and disappears.[27]

The last mentioned effect—of apparitions gradually approaching the audience—became the most famous and defining feature of the phantasmagoria (Goethe called this particular effect the "phantasmagorical manner").[28] Like the acoustic illusion of an approaching storm, the optical illusion of an approaching figure relied on a spatial interpretation of growth—an interpretation made possible by a separation of perceptual qualities from their physical origin (or "means of production," in the terms Marx would later identify with "phantasmagoria"). The approaching storm was achieved by means of a crescendo upon the standard storm instruments of drum, rain stick, and wind machine. The approaching or receding image was achieved by moving the magic lantern away from or toward the screen, respectively, so as to make the projected image grow and shrink in size.

FIGURE 5.4. E. G. Robertson, *Mémoires récréatifs, scientifiques et anecdotiques du Physicien-Aéronaute É. G. Robertson*, vol. 1 (Paris: Rignoux, 1831), frontispiece. Library of Congress, GV1545.R75 A3.

Robertson's Fantascope incorporated a mechanism for adjusting the lens with the lantern's movement, so as to maintain the image in focus.

Approaching ghosts in a dark space made the phantasmagoria an immersive experience, wherein the boundary between spectator and spectacle ceased to exist. The looming figures prompted terrified responses from spectators who drew their swords or ducked out of the ghosts' paths, much as early cinema spectators—similarly unaccustomed to the perceptual effect—reportedly reacted to film of an approaching train (figure 5.4).[29] As a Parisian spectator described one of Robertson's shows in 1800:

> The total darkness of the place, the choice of images, the astonishing magic of their truly terrifying growth . . . everything combines to strike your imagination, and to seize all your observational senses. Reason has told you well that these are mere phantoms, optical tricks devised with artistry, carried out with skill, presented with intelligence, [but] your weakened brain can only believe what it is made to see, and we believe ourselves to be transported into another world . . .[30]

Robertson contributed to the fame of phantasmagoria, but his significance has been overplayed as a result of his own careful curation of his career.

Robertson's memoir has provided a key source for historians of the phantasmagoria, leading to the form's identification with Revolutionary Paris and Robertson's particular repertoire of effects and balance between disenchantment and belief.[31]

In Germany, however, the phantasmagoria developed along different lines. Among its presenters was the German Carl Enslen—an inventor and showman now largely forgotten but in his day celebrated in the press for his mechanical genius and artistry. Enslen gained renown in the 1780s for his aerostatic sculptures—balloons shaped like animals or other figures that employed new hot-air-balloon technology to fly through the air. By 1791 he was exhibiting musical automata: concurrently in Vienna, Phylidor presented ghost shows while Enslen offered a "mechanical musical theater," featuring "life-like automata or artificial moving figures."[32] By 1796, Enslen had added ghosts to his repertoire, though he called them by the unambiguous name "optical presentation" (*optische Vorstellung*) rather than spirit appearances or phantasmagoria. At this point Enslen's shows acquired a three-act form, which he maintained on his tour of German cities until at least 1803: the first act featured performances by three musical automata; the second featured an aerostatic sculpture and an acrobatic automaton; and the third act, the optical presentation.[33]

Enslen began this optical presentation like a ghost-raiser: by securing darkness and sounding rolls of thunder, a combination that made "shudders seize the fearful," and through which "the presentiment of apparitions was stirred up."[34] He proceeded to an "optical ballet," which featured a single figure that multiplied into four, then sixteen, then reunited; these multiplied figures of genies, women, or children also danced around, moving up and down and through one another. Especially impressive to audiences, however, was the appearance of Friedrich II—the same figure Phylidor had raised in 1789. As the *Journal des Luxus und Moden* reported:

> Then a light star appeared in the distance; the star advanced, and from it developed the very life-like picture of Friedrich II, in his customary clothing and deportment, thus not as spirit. The picture becomes ever larger, comes ever nearer, until it appeared to stand life-size, close before the orchestra. The impression that this appearance made on the ground-floor and in the boxes was striking. The clapping and cheering was unceasing. When Friedrich finally withdrew back to his star, many called: stay with us! He went back to his star, but upon the loud cries of encore he had to come again a second time.[35]

Not only was Friedrich's approach immediately encored, it was also memorialized in verse:

In unabsehbar weiter Ferne	In the incalculably far distance
Aus einem lichtumstrahlten Sterne	From a shining star
Des großen Friedrichs Geist sich schuf.	The spirit of Frederick the Great appeared.[36]

Like Robertson's phantasmagoria, Enslen's optical presentation manipulated spectators' sense of space, making it seem as if his figures approached from some enormous, unknown distance. But the response was not to cower in terror but rather to welcome and celebrate the approaching figure. In addition to Friedrich II, Enslen combined the effect of approach with a military general and an allegorical "genius of peace." On one special occasion in 1796, when the Prussian king, prince, and princess attended his show, he replaced the appearance of Friedrich II with one of the current king, accompanied by the patriotic song "Heil dir im Siegeskranze," which the audience joined in singing.[37] These approaching figures provided a grand finale to his presentations, releasing spectators from their earlier, thunder-induced state of dread. They thus achieved audiovisually, in performance, the didactic purpose Phylidor and Robertson claimed for their phantasmagorias but in practice left in doubt: a triumph of reason over spirit belief.

Illusions of Motion

Phylidor, Enslen, and Robertson worked at the historically specific limits of perceptual competency and technical possibility to produce illusions of motion and to transform spectators' experience of space. But phantasmagoria showmen were not the only ones—or even the first ones—to investigate illusions of motion and their immersive, transportative potential. Attending to these aspects of phantasmagoria, rather than abstracting the term to signify first and foremost the concealment of the means of production (as has become customary following Karl Marx and Adorno), reveals a shared interest in and approach to perceptual illusions among showmen and composers, and with these a wider and more ongoing recalibration of perceptual competencies in the latter half of the eighteenth century.

The use of offstage instruments to produce spatial effects such as echoes or distant music can of course be traced to the beginnings of opera. Around the middle of the eighteenth century, however, composers began to use other techniques that—like phantasmagoria—made use of the capacity to (mis)interpret perceptual qualities spatially. Thus, distant sound could be obtained not only by relocating instruments, but also by keeping them in the orchestra and altering their tone and volume by such means as mutes (as in the case of Haydn's telescopic intermezzos, discussed in chapter 1). With this separation of perceived from actual location, sound sources could also take on the illusion of motion. In the number from *Idomeneo* (1781) discussed in the introduction, for example, Mozart produced an effect of motion toward a marching band by means of a gradual crescendo and increasingly full orchestration, from muted to unmuted instruments. As we also saw, this technique was on the edge of technical possibility: Mozart had to send from Munich to Salzburg to obtain the trumpet mutes for his crescendo, and was unable to acquire the horn mute he also desired. As his father reminded him, however, the trumpet mutes alone were "something strange and new," their novelty being essential to their effect.[38]

The combination of perceptual effects with Enlightenment symbolism and values that characterized Enslen's shows can also be found in music. In his funeral cantata for Joseph II, composed in Bonn in 1790 (the same year Phylidor introduced his new show in Vienna), Beethoven engaged the drama of superstitious belief banished by an enlightened monarch. The cantata was commissioned by Bonn's literary society on a text by local poet Severin Anton Averdonk. In the second number, the text describes the time before Joseph II's reign with an image of susceptibility to false religious beliefs: "a monster, his name Fanaticism, rose out of the depths of hell . . . and it was Night!" Beethoven prefaced this text with a fugato in D minor, the dated contrapuntal style and mode suggesting the past of darkness, superstition, and ignorance (example 5.1). In the following number, Joseph arrives to usher in a new era. Beethoven's setting, in D major, begins *pianissimo* and crescendos to the vocalist's entrance on the text "Then came Joseph, and with God's strength dragged the frenzied monster away . . . Then mankind rose into the light" (example 5.2).

The "then came Joseph" crescendo shares the key of D major and accumulating instruments with Haydn's sunrise crescendo discussed in the introduction, where the idea of transition from darkness to light is also operative. But the "then came Joseph" crescendo features notable differences: rather than a stepwise ascent, it builds triadically while a dotted motive passes between first violins and basses; and rather than an evenly paced crescendo, the *pianissimo* dynamic is maintained until the last measure of the passage, when horns, oboe, and flute finally join and a one-

EXAMPLE 5.1. Beethoven, Cantata on the Death of Emperor Joseph II, no. 2, "Ein Ungeheuer, sein Name Fanatismus," mm. 1–12

bar crescendo leads to the vocal entrance (where the dynamic falls back to *piano*). These adjustments suggest not only a change in object but also one in orientation: rather than rising from the horizon, this source of light appears in the distance and comes closer. Interpreting the passage this way, it dramatizes Joseph's coming in a manner that anticipates Enslen's optical effect—approach and illumination joining to dispel superstition with the light of reason and monarchic power.

The parallels between music and phantasmagoria demonstrate not lines of influence, but rather the similar interests and conditions—the shared strategies for producing meaning and pleasure—at work across fields we typically keep separate: music and media, auditory and visual culture, art, science, and entertainment. The common social and conceptual ground of these phenomena yielded opportunities for interaction and emergent meanings, for the training and transfer of modes of perception. For one of Phylidor's spectators in 1793, "rain, hail, and winds form simultaneously the overture and the symphony for the scene which will unfold."[39] As the phantasmagoria gained the familiarity of the magic lantern and shadow-

EXAMPLE 5.1. *(continued)*

play discussed in the previous chapter, the associative train could also run in reverse.

One artist who traversed the spaces of symphony and phantasmagoria, with influential consequences, was E. T. A. Hoffmann. Thanks to a diary entry of October 2, 1803, we know that Hoffmann read from Wiegleb's *Die natürliche Magie*, a twenty-volume compendium of artful uses of electricity, magnetism, optics, mechanics, and the like. Like other authors of natural-magic treatises, Wiegleb linked his treatment of nature's powers to the defense against fraud and superstitious beliefs:

> The knowledge of hidden natural connections will satisfy one's curiosity and thereby give him pleasure. This will also have the benefit that no one can pass off tricks of legerdemain and imposture as hitherto unknown forces of nature. And when should this be more necessary than in our time, which has produced such men as Schröpfer, Gassner, Cagliostro, the miraculous properties of magnetism, alchemical myths, and the revelation of Swedenborg's new Jerusalem?[40]

EXAMPLE 5.2. Beethoven, Cantata on the Death of Emperor Joseph II, no. 3, "Da kam Joseph," mm. 1–14

Hoffmann was also familiar with the performance culture of the physical sciences. He twice referenced Carl Enslen in his tales, with a level of detail that suggests he had witnessed Enslen's presentations (perhaps in Danzig, where Hoffmann recorded seeing automata in 1801 and where Enslen owned an estate). In Hoffmann's "The Automata" (1812), Enslen's acrobatic automaton provides a counterexample to the generalization that automata are disturbing. As a character explains,

> much depends on their workmanship and on what they do. Now there was Enslen's acrobat, one of the finest automata I have ever seen. There was a vigour about his movements which was most effective, and when he suddenly sat down on his rope, and bowed in an affable manner, he was utterly delightful. I do not suppose anyone ever experienced the gruesome feeling you speak of in looking at him.[41]

In "The Society in the Cellar" (1814), Hoffmann recalled the multiplication effect of the optical ballet to enhance the supernatural aura surrounding a

EXAMPLE 5.2. (*continued*)

mysterious character: in the candlelight, the numerous folds and creases of the fellow's coat seem to "form and divide again into many shapes or figures, like the Enslen Phantasmagoria."[42]

At the Bamberg theater, where he worked in various capacities from 1808 to 1813, Hoffmann had the opportunity to apply his knowledge of optical technologies and effects in another way. Indeed, Hoffmann credited Franz Ignaz von Holbein, who took over the theater in 1810, with initiating him "*practically* speaking into the secrets of stage effects, thereby completing the theory I had devoured from all the books I could get hold of."[43] Among the stage machinery he helped build were "the phantasmagorias" for the 1811 production of *Der standhafte Prinze* [The constant prince]. In this drama by seventeenth-century Spanish playwright Calderón de la Barca, the eponymous prince returns from the dead to guide King Alfonso and his army to victory. According to a report in the *Journal des Luxus und Moden*, Hoffmann's ghost appeared transparent and went through a wall without a sound—his "phantasmagorias" moving in three-dimensional space.[44]

Beethoven's Fifth as Phantasmagoria

That Beethoven's Fifth activated phantasmagorical associations for Hoffmann is suggested not only by the ghostly imagery throughout his discussion of the work, but also by the way in which he compares Beethoven's music with Haydn's. Haydn's symphonies, Hoffmann wrote,

> lead us into vast green meadows . . . Youths and maidens float past in circling dances . . . A life full of love, full of bliss, as before the Fall, in eternal youth; no sorrow, no pain, only a sweet melancholy yearning for the beloved object that floats in the distance out of the glow of dusk, neither approaching nor receding.[45]

By contrast (to revisit the passage discussed briefly in the introduction), Beethoven's instrumental music

> opens up the realm of the monstrous and immeasurable. Fiery beams shoot through the deep night of this realm and we become aware of giant shadows that surge back and forth, draw closer and closer in upon us, and annihilate everything in us, except the pain of infinite longing in which every desire that rushed upwards in jubilant tones sinks down and perishes; and only in this pain in which love, hope, joy are consumed, but not destroyed, and which must burst our hearts with a full-voiced chorus of all the passions, do we live on as enchanted spirit-seers [*Geisterseher*].[46]

Reading Hoffmann in the context of idealist aesthetic philosophies, Mark Evan Bonds has found the contrast between Haydn's green meadows and Beethoven's realm of the monstrous and immeasurable to express the degrees of sublimity through which art elevates the mind from the phenomenal world to the Absolute. Stephen Rumph, invoking Hegel, has identified a similar narrative of the spirit's progress from a state of nature to a higher spiritual reality. The contrast between "objects . . . in the distance neither approaching nor receding" and "giant shadows that draw closer and closer in upon us," however, invokes the key innovation of phantasmagoria—a technique that turns us into "spirit-seers." Recognizing this technique has significant implications for how we read Hoffmann's review: it suggests his imagery does not simply poeticize his philosophical and analytical arguments, but rather describes features of the music—and that his philosophical and analytical arguments were formed in response to the Fifth Symphony's phantasmagorical effects.

The Fifth Symphony begins, of course, with the force of a sound effect: a *fortissimo* pounding captured in the phrase "thus fate knocks at the door." According to Hoffmann, this beginning—plus the eight measures that follow—"determines the character of the entire piece, and for this reason the reviewer inserts it here for his readers to examine" (figure 5.5).[47]

Hoffmann singled out for mention the startling tonal ambiguity of the opening measures, and the uneasy feeling generated by arriving at a fermata on the dominant. In a review of Beethoven's *Coriolan* Overture published two years later, however, he drew attention to another musical feature also present in the opening of the Fifth: *fortissimo* unison-octaves. Here, Hoffmann made explicit the connection between loud unison-octaves, thunder, and the presentiment of apparitions. Describing as "hollow" and "eerie" the strings' sustained low Cs (which three times lead to a staccato tutti chord and spacious silence), he likened the opening of the *Coriolan* Overture to the "threatening rumble of an approaching storm," connecting the sound to the same preparatory role it played in ghost-raisings and phantasmagoria: "one fully believes that the spirit world ominously announced by subterranean thunder will draw closer in the play . . ."[48] The fermatas at the opening of the Fifth Symphony resound with a similarly "hollow" orchestration, and likewise alternate with punctuating activity and silence. The thunderous effect has the power to arouse both terror and a readiness to believe in spirits.

In the next eight measures of Hoffmann's opening music example, the famous musical motive passes from the second violins to violas to first violins, then again from seconds to violas to firsts. For the presentation of a main theme, such orchestration was highly unusual. As a rule, main themes were introduced as continuous melodies by a single instrumental voice, nearly always the first violins. To give just one example, Haydn's Symphony no. 95 (1791) shares with Beethoven's Fifth the key of C minor and unison opening. Its main theme also begins similarly, with three presentations of the same musical idea, each time higher than the last. These three presentations—along with the continuation of the theme—are all played by the first violins, with the result that we hear a continuous melodic line presented by a single, stable voice (example 5.3).

FIGURE 5.5. Beethoven, Symphony no. 5 (1808), mvt. 1, mm. 1–12, reproduced from [E. T. A. Hoffmann], "Recension," *Allgemeine musikalische Zeitung*, 12, July 4, 1810, cols. 635–636.

The opening of Beethoven's Fifth unmoors the melodic line from the single instrumental voice, dispersing it across multiple sites within the orchestra. The disorienting effect of Beethoven's orchestration can be experienced on the page, in the musical example Hoffmann provided for his readers to examine. Whereas one reads Haydn's first theme as a continuous line, following the motive in Beethoven's first theme requires the eyes to flit from one line to another, from the seconds, down to the violas, then up to the firsts. This ordering of entrances, moreover, is peculiar: the motive ascends in pitch as it moves from seconds to violas to firsts, meaning the violas play higher than the second violins (they play A-flat and G above the second violins' E-flat). What on the page forces the eyes to zigzag down and up, in the concert hall has two effects: coming between the first and second violins, the subtly different timbre of the violas reinforces, rather than masks, the passing of the motive from one instrument group to another. Performed with the traditional, antiphonal orchestral seating of the early nineteenth century, the motive passes from the right to the left side of the stage, circling the performance space with astonishing rapidity.

EXAMPLE 5.3. Haydn, Symphony no. 95 in C minor/1, mm. 1–6

Hoffmann noted Beethoven's penchant for passing short motives around the orchestra in the Fifth Symphony. Demanding a "well-trained" ensemble playing with the "utmost precision" under constant surveillance of the conductor, the technique strained orchestral capacities for coordination across space and time. Though the technique would also seem to produce "something disjointed and difficult to comprehend" for the listener, Hoffmann deemed it essential to the "arrangement of the whole . . . which hold[s] the spirit firmly in an unnameable longing."[49] Making a similar observation about Beethoven's op. 70 Piano Trios, Hoffmann elaborated with phantasmagorical imagery: "strange shapes begin a merry dance, now converging into a single point of light, now flying apart like glittering sparks, now chasing each other in infinitely varied clusters."[50] The description recalls Enslen's optical ballet, and aptly captures the effect of Beethoven's scoring, which draws the ear to focus now here, now there, following a motive whose rapid appearance, proliferation, and disappearance resembles nothing in nature so much as the dancing figures in the spirit world of phantasmagoria.

Shortly after the motivic circling, the first extended crescendo of the symphony begins (example 5.4). The start of the crescendo is also the start of the first *continuous* line in the first violins, which play an ascending version of the motive first at two-measure intervals, then at one-measure intervals, as the accompanying wind instruments gradually accumulate. The orchestration encourages us to hear the motive, now, not proliferating across the orchestra but as a single entity. Scholars frequently note the forward drive of passages such as this one, which seem to carry us along in their temporal progress. Scott Burnham writes that these measures create a sense of "driv[ing] at high speed with only the space immediately before us visible"—"there is no safe projection of the future."[51] The feeling is appropriate, but the metaphor belongs to an age of the automobile. In the world of phantasmagoria, it is not we who are hurtling into the darkness but something from the darkness hurtling toward us—drawing closer and closer with each higher and louder repetition of the motive.

The height of such phantasmagorical effect occurs with the transition from the third movement to the finale. Though Hoffmann sprinkled ghostly imagery throughout his analysis of the symphony, it was this passage that aroused for him "a horror of the extraordinary, of the fear of spirits [*Geisterfurcht*]."[52] It begins with a sustained *pianississimo* chord in the strings, and *pianissimo* timpani reducing the motive to dull, barely perceptible thuds. Then, against the continuous timpani beat, the first violins gradually wind

EXAMPLE 5.4. Beethoven, Symphony no. 5 in C minor/i, mm. 34–64

their way upward, their two-note rhythmic pattern against a three-note melodic shape cycling toward the C-major triad. Through this process, dissonance slowly transmutes into consonance, until at last the swift crescendo leads to the fourth movement with its triumphant C-major theme carried by the brass. Recalling the "then came Joseph" crescendo, we might detect a resemblance: as in the cantata, a *pianissimo* dynamic is maintained for a comparatively long time; finally, bassoons enter, then the remaining winds for a swift crescendo. The distance implied by the symphonic transition, however, far outstrips that of the cantata, and comes with an unsettling air absent from the earlier passage. This is musical phantasmagoria on a new

EXAMPLE 5.4. (*continued*)

scale: something appears at such an incalculable distance, its approach is at first barely perceptible, and we remain uncertain whether it be an object of terror or joy until it is nearly upon us (see example 2 in the introduction).

The imagery and symbolism of Beethoven's "Joseph" cantata and Enslen's optical presentations provide compelling period-specific ways in which to interpret the trajectory from darkness to light so often associated with the Fifth Symphony. With the arrival of the C-major theme, Beethoven's

EXAMPLE 5.4. (*continued*)

phantasmagoria invites the same celebratory responses as Enslen's grand finale. The return of the passage later in the movement reinforces the spectacular association. Coming after the recollection of the third-movement material, and leading once again into the C-major theme, it is at once a stroke of formal integration and an encore of the striking, beloved effect.

Hoffmann, however, made no mention of the return of the third movement and its transition to the main theme of the finale. Instead, he argued

EXAMPLE 5.4. (*continued*)

that the symphony "ris[es] in a climax right to the end . . . and irresistibly sweeps the listener into the wonderful spirit-realm of the infinite . . . Indeed for many moments after it he will be unable to emerge from the magical spirit-realm where he has been surrounded by pain and pleasure in the form of sounds."[53] Though Beethoven's phantasmagoria would seem to give every reason to experience a triumph over fear and superstition à la Enslen, Hoffmann's sense of the symphony hews closer to Robertson's shows: the spirits announced by thunder at the outset remain to the end, leaving the listener with a feeling of transport into another world, and belief in metaphysical appearances.

Beethoven's Shadow

> Hoffmann gives vent to his poetic heart in such terms as "forebodings of the realm of the spirits," "eerie feelings," "unpeaceful yearning," "terror of the unearthly—of the fear of spirits," and many more such expressions. And yet a more solid understanding of music, as opposed to the realm of the spirits, would surely have brought him closer to the point at which he wished to extend his spiritual antennae.[54]

Thus Heinrich Schenker read Hoffmann's review of Beethoven's Fifth, finding in it an atomizing, sequential description of the music peppered with subjective imagery, rather than an analysis of its organic coherence that got at its spiritual truth. The scorn with which early twentieth-century musicologists tended to regard early Romantics—their outpourings about music seeming so much poetry without rational understanding—has since been reversed: Hoffmann, along with Wackenroder and Tieck, has been installed as founder of an ideology and mode of listening appropriate to self-sufficient instrumental music. As Carl Dahlhaus influentially argued, the rhapsodic prose of these early Romantics was a symptom of their commitment to music's ineffability, its excessive and arbitrary quality being a means to let "the reader imagine what is granted the hearer of absolute music: an experience that overcomes him for an instant, but which cannot be held fast."[55] Hoffmann went further than Wackenroder and Tieck, however, in listening to "musical form as a process—rather than giving way to wild imaginings."[56] According to Dahlhaus, he "heralded a new era in musical thought" by establishing the Romantic view that "to understand music means . . . grasping the structure, the harmonic and thematic logic of a work, so as to be able to fathom its aesthetic meaning, a meaning that remains inaccessible to mere mindless enthusiasm."[57]

Hoffmann's grasp of Beethoven's Fifth Symphony helped establish it as the paradigmatic example of the unified musical work. Examining the interpretive tradition in which the symphony secured this status, Scott Burnham has found its cultural foundation in a "heroic" view of the self that flourished in German culture at the turn of the nineteenth century and continues to inform appreciations of Beethoven's music. Voiced especially by Goethe and Hegel, this concept of self emphasizes the autonomy of the individual, and the quest for a state of redemption or freedom. By unfolding the progress of a theme from struggle to victory, the Fifth Symphony (like the first movement of the *Eroica*) offers just such an experience of the self. Through identification with the thematic process, the self "experiences a culmination that is not open-ended, that leaves no loose ends. Thus the self is made to seem universal; the self and its progress is all there is."[58]

To fit the heroic narrative and deliver the experience of the self as universal, however, the Fifth Symphony must be heard to "leave no loose ends." Such hearings are common among present-day critics. Lawrence Kramer, for example, argues that the "vehement exuberance" of the finale coda balances the parallel measures of the first movement, "while the static, over-insistent C-major harmony, secure on its side of the threshold, defies the faintest peep of a minor inflection to [the] sound." William Kinderman writes that "the *presto* coda leaves all doubts about a further relapse into the minor far behind by delivering the C major 'that cannot be followed.'"[59]

Yet, in his review Hoffmann detected an unsettling lack of closure to the symphony. As he wrote, the final chords—all quarter-note blasts of C major, coming first at two-measure, than at one-measure, than again at two-measure intervals—"recall the separate strokes in the symphony's Allegro and place the listener once more in a state of tension. They act like a fire that is thought to have been put out but repeatedly bursts forth again in bright tongues of flame."[60] By recalling an earlier passage and frustrating the projection of an ending, the coda not only fails to reach unequivocal closure; it issues a positive threat of future returns.

Nor are the final chords the only place to hear such a threat. Another comes in the recurrence of the third movement within the finale. E. M. Forster recognized this in his novel *Howards End* (1910), wherein we learn the imaginative listener Helen's thoughts during a concert performance of the Fifth. For Helen, the third movement is the walk of goblins.[61] Though some might think the goblins mere "phantoms of cowardice and unbelief" easily dispelled by "one healthy human impulse," Beethoven, in Helen's interpretation, "knew better":

> The goblins really had been there. They might return—and they did . . . Beethoven chose to make all right in the end. . . . He brought back the gusts of splendor, the heroism, the youth, the magnificence of life

> and of death, and, amid vast roarings of a superhuman joy, he led his Fifth Symphony to its conclusion. But the goblins were there. They could return.[62]

Alongside the interpretive tradition in which the Fifth Symphony offers an experience of the self as universal, then, there is another, shadow tradition in which it offers an experience of the spirit world as real, alien, and ever ready to return. Hoffmann stands at the head of both traditions, earning Schenker's distancing disdain as well as his position in histories of "absolute" music, and suggesting ways in which the two are historically linked. As Terry Castle has argued, Enlightenment campaigns brought about the surmounting of ghost belief—the closure of the natural world to the supernatural—that transformed encounters with ghosts into experiences of uncanny return. Analogously, then, the symphony's progress from darkness to light and heroic closure provide the preconditions for its alternative, unsettling interpretations. In this view, Beethoven's Fifth belongs to a fully rationalized, disenchanted world, and those who perceive its ghosts are registering a sense of presence that becomes uncanny thanks to the structural boundedness, the heroic closure of the work.[63]

The historical position of Hoffmann's review, however, suggests instead that the unsettling interpretation of the symphony called for the heroic reading—that the open possibility of ghostly encounter was a precondition for the later insistence on unequivocal closure. As we have seen in this chapter, the threat of fake or unverifiable apparitions—the need to defend against charlatans and false prophets—made foreclosing the possibility of communication with the spirit world a matter of social and intellectual exigency. The experience or testimony of spirit-seeing called for a disarming explanation and philosophy. With the phantasmagoria of Beethoven's Fifth, then, to listen to surface appearances without investigating hidden connections was to court false beliefs; to hear anything other than unequivocal closure was to open oneself to deception or madness.

Scholars who place Hoffmann in the genealogy of "absolute" music have interpreted the discovery of unity in the depths of the work as the focus of his musical metaphysics. According to Dahlhaus, "for Hoffmann, the element that makes music 'cohere at its inmost center' is the same element that allows access to its metaphysical essence."[64] Abigail Chantler concurs, arguing that "for Hoffmann the physical unity of musical works was the *source* of their metaphysical meaning."[65] While later analysts would turn to underlying unity for the truth of the work, however, Hoffmann's words suggest that his analytical aim was not to discover the work's metaphysical essence, but rather to secure the status of the metaphysical appearances, of the spirit-seeing it brought forth. He argued that "a very deep examination of the inner structure of Beethoven's music will disclose its high level of rational awareness." Through deep examination, Hoffmann sought to prove that reason—not madness, illness, or mechanism—was the hidden cause of Beethoven's phantasmagoria.

Dahlhaus's suspicion of "mindless enthusiasm" and "giving way to wild imaginings"—phenomena he set in opposition to Hoffmann's structural listening—echoes the rationalist discourses that placed spirit-seeing outside the bounds of reason, and hence of knowledge. Hoffmann, however, adopted the enlightened rationalists' investigative attitude toward ghosts while foregoing their *a priori* exclusion of spirit appearances. As Adelheid Voskuhl has observed, in exploring "the 'dark' side of rational discourse: terror, fear, magic, and madness," Hoffmann and other early Romantics did not view themselves as opposing the Enlightenment, but rather as extending its modes of knowledge into as-yet-uncharted territory.[66] Hoffmann's literary oeuvre, famous for its fantastic and uncanny effects, tenders a sustained questioning of the relation between physical and spiritual worlds—a lifelong uncertainty about the legitimacy of ghosts.

Scholars have demonstrated various connections between phantasmagoria and music. Anthony Newcomb showed the rapid temporal shifts of the ghost show to be a likely source for the staging and music of the Wolf's Glen scene from Weber's *Der Freischütz*. Thomas Grey has proposed the phantasmagoria as both a background to certain works by Mendelssohn, and a general model for the "elusive status of 'musical representation'" in the nineteenth century.[67] Hoffmann's review of Beethoven's Fifth suggests placing phantasmagoria—as a set of audiovisual techniques and index of metaphysical problematics—even more centrally within the history of music. His "shadows closing in" constitute not colorful decorations upon, but rather motivations for, his aesthetic and analytical arguments. For later analysts, hewing to a rationalist line closed to the possibility of spirit-seeing, hidden unity provided a safe, suitable object of scientific investigation. For Hoffmann, it verified that listening to Beethoven's music induced visions of spiritual value. Underlying unity—connections hidden from the ear—became relevant to critical judgment, to musical experience, thanks to the problematic resemblance between projected images, wild imaginings, and metaphysical truth.

CONCLUSION
Audiovisual Returns

It is often claimed that our world has become so dominated by visual media that the ability to attend to music alone has been greatly diminished. In 1994, for instance, Leon Botstein observed a consensus within the music industry that "without a visual element or some 'interactive' component, classical music will continue to lose 'market share'"; under these conditions, "pure listening itself—without visual or linguistic elements—is at risk."[1] In 2010, Arved Ashby questioned whether one can "speak any longer of listening to music in any practical sense rather than watching-observing-listening to music," since the Internet and media convergence have produced a world in which one "rarely experiences any kind of sound without a visual key or complement." According to Ashby, the current "subservience of the musical to the visual" continues a "150-year trajectory of the 'purely musical' losing out to extramusical imagery" that began with the rise of program music in the 1830s.[2]

As of the second decade of the twenty-first century, the dependence of music on visual media seems even more decided. In 2012, a Nielsen study found that more teens listen to music on the video-streaming service YouTube than on radio, iTunes, or CD, and in 2013 the Billboard Hot 100 adjusted its formula to count YouTube plays in its weekly ranking of hit singles.[3] Numerous symphony halls have installed screens above the orchestra, onto which are projected the visual component of expanded programming selections (silent film scores or video-game music, for example), or during more canonic fare, close-up views of the musicians. Going one better, the concert hall for the New World Symphony, the New World Center, which opened in Miami in 2011, features fourteen large sail-shaped panels that surround the stage and function as both acoustic treatment and projection screens. As Alex Ross noted in a review for *The New Yorker*, "the hall is explicitly designed as much for the projection of images as for the projection of sound," the result an unprecedentedly seamless "fusion of film and live music."[4] Meanwhile, scholars have begun to speak enthusiastically of an "audiovisual turn" brought about by the new means of producing and coordinating music and images using digital media.[5]

As the preceding chapters have shown, however, not only the proliferation of lens and image-projection technologies, but also their encounters with music began far earlier than narratives of audiovisualization and histories of multimedia have allowed.[6] From the material culture and performance practices of the eighteenth century, optical technologies moved into musical discourse and habits of body and thought, helping shape the otherworldly orientation and silent, attentive listening of musical romanticism.

This audiovisual history calls for dispensing with a common assumption: that increasing focus on one sense, whether through technology use, environmental conditions, or other forms of training, necessitates a concomitant atrophying of another sense. Such is the logic behind both the historiographical model of epochal swings between visual and auditory cultures, and worries that visual media put musical listening at risk. Certainly, objects and pedagogies alike can manipulate the distribution of attention between the senses, as well as expectations as to the nature or importance of what each sense takes in. Composers of classical Hollywood cinema perfected techniques for rendering their music "unheard," while, as Brian Kane has shown, there is a long history of techniques for intensifying listening by rendering music "unseen."[7] But the prevalence of analogies between the senses and strategies of substitution attests to a much more flexible, adaptable sensorium than the notion of sense ratios, redistributing fixed perceptual resources, allows. In fact, E. T. A. Hoffmann proposed an additive rather than a competitive relation between looking and listening. As he wrote after attending a concert by composer and cellist Bernhard Heinrich Romberg in 1814, he

> wanted to see and hear [Romberg] only in his own, where he himself was the focus of attention. I say *see* and *hear* advisedly. The general desire not only to hear in a concert, but also to see, the pushing for seats in the hall, where this is possible, certainly does not arise merely from idle curiosity. One hears better when one sees. The secret relationship between light and sound is clearly demonstrated; both light and sound assume an individual form, and thus the soloist or singer himself becomes the sounding melody![8]

What Hoffmann describes here is not a mode of looking characteristic of the eighteenth century, wherein listeners watched musicians' bodily motions to clarify or reinforce the expressive character of the music.[9] Nor is it the total withdrawal from the surrounding world and tracing of musical structure that Hoffmann has been credited with modeling for future serious listeners. Instead, Hoffmann's musician seems to dissolve in the fusion of sound and moving image, and the confusion of inner and outer perception. What the New World Center accomplished through a "technological leap forward," for Hoffmann followed from the audiovisual experiences of his past.[10]

Today's "audiovisual turn," then, is also an audiovisual return—of the medial hybridities and sensory mixtures that have always characterized musical practice and experience, but have been repressed by modern aesthetics and modernist historiographies. What is new is not audiovisuality per se, nor even the audiovisual condition of music, but rather our capacity to engage with these critically and historically. Thomas Elsaesser has identified such a renovation of disciplinary acumen in cinema studies. In the twentieth century, the idea of cinema as projected, moving photography dominated histories of film, rendering marginal or ancillary to cinema history such phenomena as peep exhibition, sound recording, and music. As filmmaking, distribution, and consumption have gone digital, however, photography-projection has come to appear as one stage in a longer history of moving-image media, the inadequacy of narratives that treat it as the ultimate form forced into the open. The result is an important shift in perspective: scholars are now more ready to grant a wider range of audiovisual practices "the status of parallel or parallax cinema histories."[11]

Music historians find ourselves at a similar juncture, as the emblematic institution of our subject—the concert hall—has become a space of screens as well as musical works, the address to eyes as well as ears made plain. From the perspective of twentieth-century narratives of "the rise of instrumental music" or "the emancipation of music from language," this development appears to threaten the art, to turn musical works into musical wallpaper and absolute music into obsolete music. But it is the historiography, not the music, that is past its expiration date. Like projected photography, "pure" music now stands more clearly not as an ultimate form but as one possible, partial enframing among a wider range of mixed musical practices—practices that merit the status of music histories.[12]

Beethoven's shadow—the one his myth and music cast over subsequent generations—looms large in histories of music's aesthetic and social emancipation. But Beethoven's shadows—the giant ones closing in on us—offer a different historical picture, capturing at once the otherness of his audiovisual culture and its proximity to our own. As sources of images and music, the glass lenses, flickering flames, and acoustic instruments of the late eighteenth and early nineteenth centuries stand far removed from their early twenty-first-century counterparts. Yet, the period was, like ours, one of rapid multiplication in technological sights and sounds, of confronting the new in perceptual experiences and ways of using the body—of audiovisual culture in transition.

ACKNOWLEDGMENTS

THIS BOOK HAS EMERGED FROM CONVERSATIONS, MULTIMEDIA presentations, and written exchanges—from a mixture of auditory and visual practices—and it is a pleasure to acknowledge the many people who have played a part in the process. At the University of California-Berkeley, Mary Ann Smart shared enormous wisdom and read my work-in-progress with a keenness that continually renewed my own enthusiasm for the project. Nicholas Mathew, James Davies, and Majel Connery were critical readers and encouraging interlocutors. Richard Taruskin, Davitt Moroney, and Kate van Orden raised important points, each in their own delightful way. Students in my "Looking at Music, 1750–1850," and "Audiovisual Histories" seminars inspired me with fresh perspectives. Outside the Music Department, my fellow fellows at the Townsend Center for the Humanities and the brown-bag participants at the Center for Science, Technology, Medicine, and Society pointed me in fruitful directions. Massimo Mazzotti generously commented on material in draft. The Mellon Foundation and the American Council of Learned Societies provided funding that supported my research and made my time at Berkeley possible.

For reading parts of this book in its late stages and offering invaluable comments (some of which I am continuing to ponder), my thanks go to Thomas Grey, Thomas Patteson, Jennifer Ronyak, and Carol Vernallis. I am also grateful to everyone who contributed through pointed questions and conversations. These include Heather Hadlock, James Hepokoski, Brian Kane, Gundula Kreutzer, Richard Leppert, Ellen Lockhart, Melanie Lowe, Benjamin Piekut, Elaine Sisman, Gary Tomlinson, and Jacqueline Waeber, as well as others who responded to talks I gave at Stanford University, the American Society for Eighteenth-Century Studies, and Annual Meetings of the American Musicological Society. For insights (and good times) that have stayed with me since our days at the University of Pennsylvania, thank you to Carolyn Abbate, Lawrence Bernstein, Suzanne Bratt,

Peter Decherney, Roger Grant, Lauren Jennings, Darien Lamen, Elizabeth Mellon, Jairo Moreno, Peter Mondelli, Gavin Steingo, Leif Weatherby, Emily Zazulia, and Jason Zuzga. The Penn Humanities Forum provided a lively setting for interdisciplinary discussions and, together with the University of Pennsylvania School of Arts and Sciences, supported my archival research at the Staatsbibliothek zu Berlin, the Österreichische Nationalbibliothek, and the Wienbibliothek im Rathaus.

A special few have seen the ideas in this book through too many iterations to count and have helped me at every stage of research and writing. Emily Dolan never ceases to inspire with her sagacity and intellectual generosity. Jeffrey Kallberg remains a trusted adviser and witty reader (also sprach Loughridge) and found resources from the School of Arts and Sciences at the University of Pennsylvania to help offset the cost of the color images. John Tresch continues to set my sights on further horizons. To these admired colleagues, thank you for your continued support and guidance.

Thank you to Mark Mueller for handling my seemingly endless image-reproduction orders, and to Ursula Kwong-Brown for transcribing the music examples. Numerous collectors and curators of optical technologies kindly shared materials with me, including Richard Balzer, Erkki Huhtamo, and Laurent Mannoni. Thank you to Marta Tonegutti and Evan White at the University of Chicago Press not only for shepherding the book through publication but also for helping me hone its arguments. Many improvements to the book also resulted from the perspicacious comments of the anonymous readers, whom I thank for engaging with the manuscript so closely.

Portions of chapters 1 and 4 appear in earlier versions in the following publications: "Magnified Vision, Mediated Listening, and the 'Point of Audition' of Early Romanticism," *Eighteenth-Century Music* 10 (2013): 179–211; and "Haydn's *Creation* as an Optical Entertainment," *Journal of Musicology* 27 (2010): 9–54. Thanks to G. Henle Verlag for permission to publish excerpts from their editions of Haydn's *Il mondo della luna* and Beethoven's Fifth Piano Concerto.

A number of readers assisted with German translations. For help in this area, a particular thank you goes to Rachel Kelly, who is also a wonderful sister. Thank you to my parents for instilling in both of us a love of music and of books, and for looking forward to adding this book on music to your shelves. And last but certainly not least, thank you to Conor for taking this journey with me, and for making it so two new additions to our family—Noah and Blaise—could come along too.

NOTES

INTRODUCTION

1. Although no autograph of Symphony no. 6 survives, current scholarly consensus holds the title to be original; see Elaine Sisman, "Haydn's Solar Poetics: The *Tageszeiten* Symphonies and Enlightenment Knowledge," *Journal of the American Musicological Society* 66 (2013): 23–24. On the symbolisms of the sun, see ibid., 6–8, and Rolf Reichardt and Deborah Louise Cohen, "Light against Darkness: The Visual Representation of a Central Enlightenment Concept," *Representations* 61 (1998): 95–148.

2. On ceiling paintings depicting Apollo's sun chariot—including the Paul Troger example reproduced as figure 0.1—see Thomas Kaufman, *Painterly Enlightenment: The Art of Franz Anton Maulbertsch, 1724–1796* (Chapel Hill: University of North Carolina Press, 2005), 35–37; on the Carpoforo Tencalla ceiling paintings at Esterházy's Eisenstadt palace and their potential connections to Haydn's Symphony no. 6, see Sisman, "Haydn's Solar Poetics," 29–32; Thomas Tolley, *Painting the Cannon's Roar: Music, the Visual Arts, and the Rise of an Attentive Public in the Age of Haydn, c. 1750 to c. 1810* (Burlington, VT: Ashgate, 2001), 81–83.

3. "Die Musik schliesst dem Menschen ein unbekanntes Reich auf; eine Welt, die nichts gemein hat mit der äussern Sinnenwelt . . ." [E. T. A. Hoffmann], "Recension," *Allgemeine musikalische Zeitung* 12, July 4, 1810, col. 631; trans. E. T. A. Hoffmann, *E. T. A. Hoffmann's Musical Writings*, ed. David Charlton, trans. Martyn Clarke (Cambridge: Cambridge University Press, 1989), 236.

4. Scholarship establishing this narrative around Hoffmann's review includes Carl Dahlhaus, *The Idea of Absolute Music* (Chicago: University of Chicago Press, 1989); Lydia Goehr, *The Imaginary Museum of Musical Works: An Essay in the Philosophy of Music* (New York: Oxford University Press, 1992); Mark Evan Bonds, "Idealism and the Aesthetics of Instrumental Music at the Turn of the Nineteenth Century," *Journal of the American Musicological Society* 50 (1997): 387–420. Mary Hunter has revised the notion that an ideal of faithful, self-obliterating performance was characteristic of early Romanticism, instead finding notions of the performer's self-transformation and union with the composer in creative activity; "'To Play as if from the Soul of the Composer': The Idea of the Performer in Early Romantic Aesthetics," *Journal of the American Musicological Society* 58 (2005): 357–398.

5. ". . . wir werden Riesenschatten gewahr, die . . . enger und enger uns einschließen." Hoffmann, "Recension," col. 633.

6. Marshall McLuhan, *Gutenberg Galaxy: The Making of Typographic Man* (Toronto: University of Toronto Press, 1962), 68.

7. Emily Thompson, *The Soundscape of Modernity: Architectural Acoustics and the Culture of Listening in America, 1900–1933* (Cambridge, MA: MIT Press, 2002); Jonathan Sterne, *The Audible Past: Cultural Origins of Sound Reproduction* (Durham, NC: Duke University Press, 2003); *Hearing Cultures: Essays on Sound, Listening, and Modernity*, ed. Veit Erlmann (New York: Berg, 2004).

8. Jonathan Crary, *Techniques of the Observer: On Vision and Modernity in the Nineteenth Century* (Cambridge, MA: MIT Press, 1990); Sterne, *Audible Past*; Friedrich Kittler, *Gramophone, Film,*

Typewriter, trans. Geoffrey Winthrop-Young (Stanford, CA: Stanford University Press, 1999); Lutz Koepnick and Nora M. Alter, eds., *Sound Matters: Essays on the Acoustics of Modern German Culture* (New York: Berghahn Books, 2004); Caroline A. Jones, *Eyesight Alone: Clement Greenberg's Modernism and the Bureaucratization of the Senses* (Chicago: University of Chicago Press, 2005).

9. Seminal scholarship here includes Mary Ann Smart, *Mimomania: Music and Gesture in Nineteenth-Century Opera* (Berkeley: University of California Press, 2004); Nicholas Cook, *Analysing Musical Multimedia* (Oxford: Oxford University Press, 1998); Michel Chion, *Audio-Vision: Sound on Screen*, ed. and trans. Claudia Gorbman (New York: Columbia University Press, 1994); Rick Altman, *Silent Film Sound* (New York: Columbia University Press, 2007).

10. W. J. T. Mitchell, "Showing Seeing: A Critique of Visual Culture," *Journal of Visual Culture* 1 (2002): 166.

11. Anselm Gerhard, *The Urbanization of Opera: Music Theater in Paris in the Nineteenth Century* (Chicago: University of Chicago Press, 1998), 6.

12. Sarah Hibberd, "*Le Naufrage de la Méduse* and Operatic Spectacle in 1830s Paris," *19th-Century Music* 36 (2013): 248–263.

13. John Tresch, *The Romantic Machine: Utopian Science and Technology After Napoleon* (Chicago: University of Chicago Press, 2012), 135. See also his "The Prophet and the Pendulum: Sensational Science and Audiovisual Phantasmagoria around 1848," *Grey Room* 43 (2011): 16–41.

14. Holly Watkins, *Metaphors of Depth in German Musical Thought: From E. T. A. Hoffmann to Arnold Schoenberg* (New York: Cambridge University Press, 2011), 22–50; Mark Evan Bonds, *Music as Thought: Listening to the Symphony in the Age of Beethoven* (Princeton, NJ: Princeton University Press, 2006).

15. Emily Dolan, *The Orchestral Revolution: Haydn and the Technologies of Timbre* (Cambridge: Cambridge University Press, 2013), 88.

16. See the concluding chapter for discussion of this historiographical myth.

17. Alexander Gottlieb Baumgarten, *Mediationes philosophicae de nonnullis ad poema pertinentibus* (Halle: J. Grunerti, 1734), cxv–cxvi, cited in Michael McKeon, "Mediation as Primal Word: The Arts, the Sciences, and the Origins of the Aesthetic," in *This Is Enlightenment*, ed. Clifford Siskin and William Warner (Chicago: University of Chicago Press, 2010), 384–412.

18. Johann Gottfried Herder, "*Kalligone* (1800)," trans. in *Musical Aesthetics: A Historical Reader*, vol. 2: *The Nineteenth Century*, ed. Edward A. Lippman (New York: Pendragon Press, 1988), 41. Gotthold Ephraim Lessing's *Laokoon: Oder, Über die Grenzen der Malerei und Poesie* (1766) is typically credited with differentiating the arts on the basis of space and time. Because he reasoned from the nature of signs rather than the senses, his theory encourages formalist approaches to the arts over histories of listening and looking. For studies that take Lessing's *Laokoon* as a point of departure for tracing medium specificity versus synthesis in the arts, see Daniel Albright, *Untwisting the Serpent: Modernism in Music, Literature, and the Other Arts* (Chicago: University of Chicago Press, 2000); Simon Shaw-Miller, *Visible Deeds of Music: Art and Music from Wagner to Cage* (New Haven, CT: Yale University Press, 2002); Holly Rogers, *Sounding the Gallery: Video and the Rise of Art-Music* (New York: Oxford University Press, 2013), 46–81.

19. David Howes discusses the eighteenth-century emergence of a fine arts taxonomy founded on the dualism of vision and hearing; smell, taste, and touch were considered too lowly to be significant to the fine arts, while theater and dance were sidelined due to their hybrid character. See his "Hearing Scents, Tasting Sights: Towards a Cross-Cultural Multimodal Theory of Aesthetics," in *Art and the Senses*, ed. Francesca Bacci and David Melcher (New York: Oxford University Press, 2013), 168.

20. Brian Kane points to an earlier history for the "separation and coordination of the senses" in the philosophical writings of John Locke. Brian Kane, *Sound Unseen: Acousmatic Sound in Theory and Practice* (New York: Oxford University Press, 2014), 86. The multiplication effect of the separation of the senses on the audiovisual is especially apparent in cinema, where the unity of the auditory and visual, having been severed in the recording process, is not given. See Chion, *Audio-Vision*; Rick Altman, "Moving Lips: Cinema as Ventriloquism," *Yale French Studies* 60 (1980): 67–79.

21. Johann Gottfried Herder, "*Kalligone* (1800)", trans. in *Musical Aesthetics: The Nineteenth Century*, ed. Edward Lippman, 41.

22. Wilhelm Heinrich Wackenroder and Ludwig Tieck, *Phantasien über die Kunst* (Hamburg, 1799), 72, trans. (here slightly altered) in Wilhelm Heinrich Wackenroder, *Wilhelm Heinrich Wackenroder's Confessions and Fantasies*, trans. Mary Hurst Schubert (University Park: Pennsylvania State University Press, 1971), 192.

23. Mark Evan Bonds cites the first part of this passage as exemplifying early Romantics' application of idealist philosophy to music, while disregarding the "and yet." Bonds, *Music as Thought*, 23.

24. Wackenroder and Tieck, *Phantasien über die Kunst*, 96. Richard Will discusses the qualifying conjunctions in this as well as the afore-cited Wackenroder passage from the perspective of the relationship between music and language in *The Characteristic Symphony in the Age of Haydn and Beethoven* (Cambridge: Cambridge University Press, 2002), 14–17.

25. Walter Benjamin used the term "perceptual world" (*Merkwelt*) to refer not simply to phenomena in the environment at a given point in history, but also to their apperception by what he called the "mimetic faculty," the human capacity to produce or recognize similarities. I similarly use the term to capture not only phenomena available to be perceived, but also ways of perceiving. See Walter Benjamin, "Doctrine of the Similar (1933)," trans. Knut Tarnowski, *New German Critique* 17 (1979): 65–69.

26. Robert Jütte, *A History of the Senses: From Antiquity to Cyberspace* (Malden, MA: Polity, 2005), 8.

27. Katherine Hayles, *How We Became Posthuman* (Chicago: University of Chicago Press, 1999), 206–207.

28. Jean-Jacques Rousseau, *Dictionnaire de musique*, in *Music and Aesthetics in the Eighteenth and Early-Nineteenth Centuries*, ed. Peter le Huray and James Day (New York: Cambridge University Press, 1988), 86. See chapter 2 for further discussion of Rousseau's aesthetic theory, and chapter 3 on the audiovisuality of poetry for late eighteenth-century readers.

29. For further exemplification and analysis of the continuity between concert life and spectacle in early nineteenth-century Vienna, see Nicholas Mathew, *Political Beethoven* (Cambridge: Cambridge University Press, 2013), 79–101.

30. Johann Mattheson, *Grundlage einer Ehren-Pforte* (Hamburg: Mattheson, 1740), 355, trans. (modified) in Steven Zohn, "Images of Telemann: Narratives of Reception in the Composer's Anecdote, 1750–1830," *Journal of Musicology* 21 (2004): 480; G. W. F. Hegel, *Aesthetics: Lectures on Fine Art*, vol. 2, trans. T. M. Knox (London: Oxford University Press, 1975), 902.

31. David Gramit, *Cultivating Music: The Aspirations, Interests, and Limits of German Musical Culture, 1770–1848* (Berkeley: University of California Press, 2002); Celia Applegate, *Bach in Berlin: Nation and Culture in Mendelssohn's Revival of the St. Matthew Passion* (Ithaca, NY: Cornell University Press, 2005).

32. Rita Felski, "'Context Stinks!'" *New Literary History* 42 (2011): 590.

33. On the innovations and influence of *opéra comique*, see Edward J. Dent, *The Rise of Romantic Opera* (New York: Cambridge University Press, 1976), 24; *The Age of Beethoven, 1790–1830*, ed. Gerald Abraham (New York: Oxford University Press, 1982); David Charlton, *French Opera, 1730–1830: Meaning and Media* (Brookfield: Ashgate, 2000).

34. James H. Johnson, *Listening in Paris: A Cultural History* (Berkeley: University of California Press, 1995), 88, 225.

35. Letters of December 1780 in Wolfgang Amadeus Mozart, *The Letters of Mozart and His Family*, vol. 2, trans. and ed. Emily Anderson (New York: St. Martin's Press, 1966), 682–683.

36. Anne Friedberg, *Window Shopping: Cinema and the Postmodern* (Berkeley: University of California Press, 1994), 28.

37. Thomas Elsaesser, "Early Film History and Multi-Media: An Archaeology of Possible Futures?" in *New Media, Old Media: A History and Theory Reader*, ed. Wendy Hui Kyong Chung and Thomas Keenan (New York: Routledge Press, 2006), 17.

38. On the importance of optical media to Kant and other idealist philosophers, see Stefan An-

driopoulos, *Ghostly Apparitions: German Idealism, the Gothic Novel, and Optical Media* (New York: Zone Books, 2013).

39. Stephan Oetterman, *The Panorama: History of a Mass Medium*, trans. Deborah Lucas Schneider (New York: Zone Books, 1997), is the main study of the impact of the panorama on perception. Anno Mungen has shown the panoramas and tableaux vivants to be central to the development of "picture music" in the nineteenth century. Mungen, *"BilderMusik": Panorama, Tableux vivants und Lichtbilder als multimediale Darstellungsformen in Theater- und Musikaufführungen vom 19. bis zum frühen 20. Jahrhundert*, 2 vols. (Remscheid: Gardez! Verlag, 2006). Other studies relating the panorama to music are Gerhard, *Urbanization of Opera*; Tresch, *Romantic Machine*; Hibberd, "*Le Naufrage de la Méduse* and Operatic Spectacle in 1830s Paris"; and Ulrich Schmitt, *Revolution im Konzertsaal: zur Beethoven-Rezeption im 19. Jahrhundert* (New York: Schott, 1990). Gerhard, Tresch, and Hibberd make connections to the stage spectacle of grand opera; Schmitt argues more generally for the panorama shifting (musical) perception away from detail and toward the whole.

40. ". . . leben wir fort und sind entzückte Geisterseher." Hoffmann, "Recension," col. 633.

41. Peter Franklin, *Seeing Through Music: Gender and Modernism in Classic Hollywood Film Scores* (New York: Oxford University Press, 2011), 90.

42. Walter Benjamin, "On the Concept of History," in *Benjamin: Selected Writings*, vol. 4, ed. Howard Eiland and Michael W. Jennings (Cambridge, MA: Harvard University Press, 2003), 397.

43. Carl Dahlhaus, *Nineteenth-Century Music*, trans. J. Bradford Robinson (Berkeley: University of California Press, 1989), 91.

CHAPTER ONE

1. Franco Piva, notes to *Galuppi: Il mondo della luna* (Bongiovanni B00000I597, 1997), 30–31 (translations modified).

2. *Galuppi: Il mondo della luna*, 32. Baldassare Galuppi, who was the first to compose Goldoni's *Il mondo della luna* with his version for Venice in 1750, set the scene according to this design.

3. Jonathan Sterne, *The Audible Past: Cultural Origins of Sound Reproduction* (Durham, NC: Duke University Press, 2003), 142.

4. On the history and meaning of the term *prosthesis*, to which the man-machine interface and the twin implications of extension and compensation are central, see Tim Armstrong, *Modernism, Technology, and the Body: A Cultural Study* (New York: Cambridge University Press, 1998), 77–84.

5. Robert Hooke, *Micrographia: or, Some Physiological Descriptions of Minute Bodies Made by Magnifying Glasses with Observations and Inquiries Thereupon* (London: J. Martyn and J. Allestry, 1665), preface.

6. James Bennett, "The Mechanics' Philosophy and the Mechanical Philosophy," *History of Science* 24 (1986): 1–28. See also Antoni Malet, "Early Conceptualizations of the Telescope as an Optical Instrument," *Early Science and Medicine* 10 (2005): 237–262; and Steven Shapin and Simon Schaffer, *Leviathan and the Air-Pump: Hobbes, Boyle, and the Experimental Life* (Princeton, NJ: Princeton University Press, 2011), esp. 36–38.

7. Seminal here is Marshall McLuhan, *Understanding Media: The Extensions of Man* (New York: McGraw Hill, 1964).

8. See Thomas L. Hankins and Robert J. Silverman, *Instruments and the Imagination* (Princeton, NJ: Princeton University Press, 1995), 4. Albert Van Helden and Thomas L. Hankins, "Introduction: Instruments in the History of Science," *Osiris* 9 (1994): 3.

9. Claude Adrien Helvétius, *De l'esprit; or Essays on the Mind, and Its Several Faculties*, trans. William Mudford (London: M. Jones, 1807), 11–12. Kant refers to Helvétius's telling of the story in his 1764 *Versuch über die Krankheiten des Kopfes* [Essay on the diseases of the mind], and in his later lectures on anthropology; see Immanuel Kant, *Anthropology from a Pragmatic Point of View*, ed. and trans. Robert B. Louden (Cambridge: Cambridge University Press, 2006), 72.

10. Catherine Wilson, "Visual Surface and Visual Symbol: The Microscope and the Occult in Early Modern Europe," *Journal of the History of Ideas* 49 (1988): 101.

11. See Michael Brago, "Haydn, Goldoni, and *Il mondo della luna*," *Eighteenth-Century Studies* 17 (1984): 310.

12. Trans. in Pierpaolo Polzonetti, *Italian Opera in the Age of the American Revolution* (Cambridge: Cambridge University Press, 2011), 53.

13. Wilson, "Visual Surface and Visual Symbol," 101.

14. Patrick Singy, "Huber's Eyes: The Art of Scientific Observation Before the Emergence of Positivism," *Representations* 96 (2006): 54–75.

15. Alexander Pope, *Essay on Man* (1732–1744), quoted in ibid., 59.

16. Polzonetti, *Italian Opera in the Age of the American Revolution*, 101.

17. Ibid., 102.

18. Hans Blumenberg, *The Legitimacy of the Modern Age*, trans. Robert M. Wallace (Cambridge, MA: MIT Press, 1985), 373.

19. Hooke, *Micrographia*, 186.

20. *Galuppi: Il mondo della luna*, 33.

21. For an example of microscopic looking within from Cosmus Conrad Cuno's *Observationes durch dessen verfertigte Microscopia* (Augsburg: Samuel Fincke, 1734), see Deirdre Loughridge, "Magnified Vision, Mediated Listening, and the 'Point of Audition' of Early Romanticism," *Eighteenth-Century Music* 10 (2013): 189–190.

22. Jan Golinski, "Barometers of Change: Meteorological Instruments as Machines of Enlightenment," in *The Sciences in Enlightened Europe*, ed. William Clark, Jan Golinski, and Simon Schaffer (Chicago: University of Chicago Press, 1999), 72.

23. Benjamin Martin, *The Young Gentleman and Lady's Philosophy: In a Continued Survey of the Works of Nature and Art by Way of Dialogue*, vol. 3 (London: W. Owen, 1782), 33.

24. Cuno, *Observationes durch dessen verfertigte Microscopia.*

25. Immanuel Kant, *Observations on the Feeling of the Beautiful and Sublime and other Writings*, ed. and trans. Patrick Frierson and Paul Guyer (Cambridge: Cambridge University Press, 2011), 36–37.

26. Brago, "Haydn, Goldoni, and *Il mondo della luna*," 310–315.

27. Polzonetti, *Italian Opera in the Age of the American Revolution*, 102.

28. Sisman, "Haydn's Solar Poetics," 42–43.

29. On Herschel, see Richard Holmes, *The Age of Wonder: How the Romantic Generation Discovered the Beauty and Terror of Science* (New York: Pantheon Books, 2008). On Haydn's visit to Herschel's telescope, see Laurence Joyce, "Haydn and Salomon, and Their Visit to Herschel's Telescope at Slough, June 1792," *Haydn Studien* 8 (2000–2004): 289–292; and Sisman, "Haydn's Solar Poetics," 45.

30. Polzonetti, *Italian Opera in the Age of the American Revolution*, 89.

31. "besser auszudrücken," Leopold Mozart, *Versuch einer gründlichen Violinschule* (Wien: C. Stephenson, 1922), 52; trans. in Leopold Mozart, *A Treatise on the Fundamental Principles of Violin Playing*, trans. Editha Knocker (New York: Oxford University Press, 1951), 89. "lebhafter auszudrücken," Johann Joachim Quantz, *Versuch einer Anweisung, die Flöte traversiere zu spielen: Reprint der Ausgabe Berlin 1752* (München: Deutscher Taschenbuch Verlag, 1992), 203; trans. in Johann Joachim Quantz, *On Playing the Flute*, trans. Edward Reilly (New York: Schirmer Books, 1975), 233.

32. For further discussion of Haydn's use of muted violins, see Deirdre Loughridge, "*Technologies of the Invisible: Optical Instruments and Musical Romanticism*" (PhD diss., University of Pennsylvania, 2011), 14–22.

33. In addition to being widely performed, *Zémire et Azor* was owned in printed score by Mozart and directed by Haydn in Italian translation in 1782 and by E. T. A. Hoffmann in a German version in 1813. See Daniel Heartz, *Music in European Capitals: The Galant Style, 1720–1780* (New York: Norton, 2003), 791; H. C. Robbins Landon, *Haydn: Chronicle and Works*, 5 vols. (Bloomington: Indiana University Press, 1976–1980), 2:61; Abigail Chantler, *E. T. A. Hoffmann's Musical Aesthetics* (Aldershot: Ashgate, 2006), 161.

34. Jean-François Marmontel and André-Ernest-Modeste Grétry, *Zemire et Azor: comédie-ballet, en vers, et en quatre actes* (Paris: Vente, 1771), 48.

35. Stefano Castelvecchi, "From Nina to Nina: Psychodrama, Absorption, and Sentiment in

the 1780s," *Cambridge Opera Journal* 8 (1996): 91–112; and Downing A. Thomas, *Aesthetics of Opera in the Ancien Régime, 1647–1785* (Cambridge: Cambridge University Press, 2002), 260–262.

36. Castelvecchi, "From Nina to Nina," 98.

37. David Charlton, *Grétry and the Growth of Opéra-Comique* (Cambridge: Cambridge University Press, 1986), 103; Thomas, *Aesthetics of Opera*, 260; Heartz, *Music in European Capitals*, 791. Grétry, however, attributed the contours of the melody to the declamation of the text "Ah! Laissez-moi, laissez-moi"; see Heartz, *Music in European Capitals*, 790–791.

38. David Buch aligns the instrumentation of the trio with that of "otherworldly scenes of supernatural operas" and identifies a number of later operas employing the same instrumentation for scenes of enchantment in *Magic Flutes & Enchanted Forests: The Supernatural in Eighteenth-Century Musical Theater* (Chicago: University of Chicago Press, 2008), 135. Thomas emphasizes distance in *Aesthetics of Opera*, 260–261.

39. Charlton, *Grétry*, 103–104.

40. "Avec deux aunes de moire d'argent, pour imiter la glace du trumeau, et deux aunes de gaze claire et transparent, je [le décorateur] appris à produire l'une des plus agréables illusions du théâtre." Jean-François Marmontel, *Mémoires*, vol. 3 (Paris: Xhrouet, 1804), 132–133. Translation slightly modified from *Memoirs of Marmontel*, trans. Hugh Murray, vol. 3–4 (Edinburgh: Mundell, Doig, and Stevenson, 1808), 136.

41. Gösta M. Bergman, *Lighting in the Theatre* (Stockholm: Almqvist & Wiksell International, 1977), 128–129.

42. ". . . on ne pourra jamais affez louer le Trio en *sourdine* du pere & de ses deux filles qui paroissent dans le tableau magique." "Comédie Italienne," *Mercure de France* (January 1772), 163; trans. Thomas, in *Aesthetics of Opera*, 260. *Journal helvétique; ou Annales littéraries et politiques de l'Europe* (January 1772), 42, also singled out the "trio en sourdine" for praise.

43. Caryl Clark, "Haydn in the Theater: The Operas," in *The Cambridge Companion to Haydn*, ed. Caryl Clark (Cambridge: Cambridge University Press, 2005), 185.

44. Landon discusses the significance of Haydn's key scheme in *Haydn: Chronicle and Works*, 2:525.

45. A German translation of *Il mondo della luna*, published for Baldassare Galuppi's setting of 1750, provides a stage direction for the "illuminated machine" that suggests how it may have been executed: "Man sieht eine erleuchtete Maschiene, die sich über der Spitze des Tubus hin und her bewegt, und auf welcher es Figuren hat, wie sie gewöhnlich auf den Gläsern der *laterna magica* sind." [One sees an illuminated machine that moves back and forth above the end of the (optical) tube, and on which are figures like those usually on the glass (slides) of a magic lantern.] *Die Welt im Monde* (Oels: Samuel Gottlieb Ludwig, [c. 1750]), 8.

46. Scholarship that has identified static and repetitive qualities with the visual aesthetic of theatrical tableaux includes Elisabeth Le Guin, *Boccherini's Body: An Essay in Carnal Musicology* (Berkeley: University of California Press, 2006), and Nancy November, "Instrumental Arias or Sonic Tableaux: 'Voice' in Haydn's String Quartets Opp. 9 and 17," *Music & Letters* 89 (2008): 346–372.

47. Ludwig Tieck, *William Lovell: Ein Briefroman aus den Jahren 1793–1796* (Darmstadt: Schriftenreihe Agora, 1961), 171.

48. E. T. A. Hoffmann, "Master Flea," in *The Golden Pot and Other Tales*, trans. Ritchie Robertson (New York: Oxford University Press, 2000), 239–375.

49. On the body as "soul made visible," see Klaus Dörner, *Madmen and the Bourgeoisie* (Oxford: Basil Blackwell, 1981), 188.

50. See Arthur Loesser, *Men, Women, and Pianos: A Social History* (New York: Simon and Schuster, 1954); Annette Richards, *The Free Fantasia and the Musical Picturesque* (Cambridge: Cambridge University Press, 2001); Samuel Breene, "The Instrumental Body in the Age of Mozart: Science, Aesthetics and Performances of the Self," *Early Music* 42 (2014): 231–247.

51. Richards, *Free Fantasia*, 76.

52. Peter Schleuning, "Die Fantasiermaschine: Ein Beitrag zur Geschichte der Stilwende um

1750," *Archiv für Musikwissenschaft* 27 (1970): 200; partially translated in Richards, *The Free Fantasia*, 175.

53. Immanuel Kant, *Anthropology from a Pragmatic Point of View*, 23–24.

54. Ibid., 24–25.

55. Carl Friedrich Pockels (ed.), *Denkwürdigkeiten zur Bereicherung der Erfahrungsseelenlehre und Characterkunde*, vol. 1 (Halle: Rengerschen Buchhandlung, 1794), 148–149, trans. Theodore Ziolkowski, *German Romanticism and Its Institutions* (Princeton, NJ: Princeton University Press, 1992), 145.

56. "Die meisten Lebenden hatten für mich nur das Interesse, das die Todten für den Anatomen haben, und es schmerzte mich oft, dass man nicht einmal eine recht bedeutende Seele mit der Zanger fassen und unter das Mikroskop bringen könnte." Friedrich Rochlitz, "Der Besuch im Irrenhause," *Allgemeine musikalische Zeitung* 6/39 (June 27, 1804), col. 645; trans. in Richards, *Free Fantasia*, 145.

57. Rochlitz, "Der Besuch," col. 645. Francesca Brittan describes later developments in the association between magnifying instruments and artistic inspiration in "On Microscopic Hearing: Fairy Magic, Natural Science, and the *Scherzo Fantastique*," *Journal of the American Musicological Society* 64 (2011): 527–600.

58. Rochlitz, "Der Besuch," col. 648.

59. Ibid., col. 651; trans. in Richards, *Free Fantasia*, 147.

60. John T. Hamilton, *Music, Madness, and the Unworking of Language* (New York: Columbia University Press, 2008), 179.

61. "Ob es also nur ganz unwillkührliche Bewegung war, wie etwa das öftere, schnelle Blinken mit den Augen bey manchen andern heftigen Menschen im Affekt; oder ob seiner, dann erhitzten Phantasie Gestalten vorschwebten, zu denen er wirklich in einer neuen Sprache zu sprechen und verständlich zu sprechen glaubte; oder such ob er durch das öftere Zusammendenken von 'Sprache' und 'Musik'—(z. B. Musik, Sprache des Herzens, ohne Worte u. dgl.) beydes, wenn er entzündet wurde, erst im Begriff, hernach in der Ausführung vermischte, oder wie es sonst damit war: das weiss ich nicht." Rochlitz, "Der Besuch," col. 706.

62. Richards, *Free Fantasia*, 183.

63. John Russell, *A Tour in Germany, and Some of the Southern Provinces of the Austrian Empire, in the Years 1820, 1821, 1822*, vol. 2 (Edinburgh: Archibald Constable, 1825), 276–277.

64. Ibid., 275–276.

65. See, for example, the account of Mrs. Grillparzer eavesdropping on Beethoven when the two were neighbors at Heiligenstadt in Alexander Wheelock Thayer, *Thayer's Life of Beethoven*, rev. and ed. Elliot Forbes (Princeton, NJ: Princeton University Press, 1967), 441–442.

66. Tia DeNora, *Beethoven and the Construction of Genius: Musical Politics in Vienna, 1792–1803* (Berkeley: University of California Press, 1997), 145–146.

67. K. M. Knittel, "Pilgrimages to Beethoven: Reminiscences by His Contemporaries," *Music & Letters* 84 (2003): 19–54.

68. See, for example, Joseph Kerman, "Beethoven Quartet Audiences: Actual, Potential, Ideal," in *The Beethoven Quartet Companion*, ed. Robert Winter and Robert Martin (Berkeley: University of California Press, 1994), 26–27, and Richard Taruskin, *The Oxford History of Western Music*, vol. 2: *Music in the Seventeenth and Eighteenth Centuries* (Oxford: Oxford University Press, 2005), 682–684.

69. In 1809 Beethoven also wrote cadenzas for his earlier piano concertos, which had been published under the assumption that soloists would improvise. On Beethoven's cadenzas, see Robin Stowell, *Beethoven: Violin Concerto* (Cambridge: Cambridge University Press, 1998), 91, and Stephen C. Rumph, *Beethoven after Napoleon: Political Romanticism in the Late Works* (Berkeley: University of California Press, 2004), 103–105.

70. Leon Plantinga, *Beethoven's Concertos: History, Style, Performance* (New York: Norton, 1999), 275–276, and Rumph, *Beethoven after Napoleon*, 103–105.

71. Joseph Kerman, *Concerto Conversations* (Cambridge, MA: Harvard University Press, 1999), 24.

72. William Kinderman, *Beethoven*, 2nd ed. (Oxford: Oxford University Press, 2009), 135; Plantinga, *Beethoven's Concertos*, 267.

73. Simon P. Keefe, "The Concerto from Mozart to Beethoven: Aesthetic and Stylistic Perspectives," in *The Cambridge Companion to the Concerto*, ed. Simon P. Keefe (Cambridge: Cambridge University Press, 2005), 91.

74. See Owen Jander, *Beethoven's "Orpheus" Concerto: The Fourth Piano Concerto in Its Cultural Context* (Hillsdale, NY: Pendragon, 2009); and Michael P. Steinberg, *Listening to Reason: Culture, Subjectivity, and Nineteenth-Century Music* (Princeton, NJ: Princeton University Press, 2006), 63–67.

75. Carl Czerny, *Die Kunst des Vortrags der älteren und neueren Klavierkompositionen* (Vienna: A. Diabelli u. Comp., 1846), trans. Carl Czerny, *On the Proper Performance of All Beethoven's Works for the Piano*, ed. Paul Badura-Skoda (Vienna: Universal, 1970), 113.

76. Ibid.

77. James Webster, "The Rhetoric of Improvisation in Haydn's Keyboard Music," in *Haydn and the Performance of Rhetoric*, ed. Tom Beghin and Sander M. Goldberg (Chicago: University of Chicago Press, 2007), 208; Rumph, *Beethoven after Napoleon*, 105.

78. On reverberant tone, see Edward A. Lippman, "The Tonal Ideal of Romanticism," in *The Philosophy & Aesthetics of Music* (Lincoln: University of Nebraska Press, 1999), 123–135.

79. Plantinga, *Beethoven's Concertos*, 144.

80. See the chapters on pedaling from early nineteenth-century piano treatises translated in the appendix to David Rowland, *A History of Pianoforte Pedalling* (Cambridge: Cambridge University Press, 1993).

81. Quoted in Keefe, "The Concerto from Mozart to Beethoven," 91.

82. John Tresch, "Estrangement of Vision: Edgar Allen Poe's Optics," in *Observing Nature—Representing Experience: The Osmotic Dynamics of Romanticism, 1800–1850*, ed. Erna Fiorenti (Berlin: Reimer, 2007), 145.

CHAPTER TWO

1. Barbara Maria Stafford, *Body Criticism: Imaging the Unseen in Enlightenment Art and Medicine* (Cambridge, MA: MIT Press, 1993), 336.

2. Jonathan Crary, "Géricault, the Panorama, and Sites of Reality in the Early Nineteenth Century," *Grey Room* 9 (2002): 8–9.

3. This formulation of listening toward or in anticipation of looking owes to (and contrasts with) Michal Grover-Friedlander's characterization of Italian-style opera in terms of "always listening in anticipation of, or listening toward, a place where one knows beautiful singing will take place." Michal Grover-Friedlander, *Vocal Apparitions: The Attraction of Cinema to Opera* (Princeton, NJ: Princeton University Press, 2005), 3.

4. In addition to the tradition of natural magic, cultivated especially by seventeenth-century Jesuits, the concept for the apparatus has been traced to the linear-perspective machines employed by sixteenth- and seventeenth-century artists to help map three-dimensional spaces onto two-dimensional surfaces. See Erkki Huhtamo, "Toward a History of Peep Practice," in *A Companion to Early Cinema*, ed. André Gaudreault, Nicolas Dulac, and Santiago Hidalgo (Hoboken, NJ: John Wiley & Sons, 2012), 32–51. On linear-perspective machines, see Anne Friedberg, *The Virtual Window* (Cambridge, MA: MIT University Press, 2006), 38–42.

5. On Engelbrecht's *Kullisentheater*, see Georg Füsslin, *Der Guckkasten: Einblick, Durchblick, Ausblick* (Stuttgart: Füsslin, 1995), 46–64; and Frances Terpak, "Perspective Theaters," in *Devices of Wonder: From the World in a Box to Images on a Screen*, ed. Barbara Maria Stafford and Frances Terpak (Los Angeles: Getty Research Institute, 2001), 336–343.

6. David Robinson connects Engelbrecht's opera scenes to the stage designs of Giuseppe Galli Bibbiena, whose collection of perspectival, architectural illustrations was published in Augsburg in the 1740s; Robinson also points to Engelbrecht's publication of Pietro Righini's operatic stage designs as a source for his paper theater. See David Robinson, "Augsburg Peepshows," *Print Quarterly* 5 (1988): 191.

7. "Raritäten-Kasten, ist ein Kasten in welchem diese oder jene alte oder neue Geschichte im kleinen und durch darzu verfertigtes Puppenwerck, so gezogen werden kan, vorgestellet wird. Es pflegen gemeine Leute, so mehrentheils Italiäner von Geburth, mit solchen Kasten die Messen in Deutschland zu besuchen, auf den Gassen herum zu lauffen und durch ein erbärmliches Geschrey: Schöne Rarität! Schöne Spielwerck! Liebhaber an sich zu locken, die vors Geld hinein sehen. Weil nun solche Dinge mehr vor Kinder als erwachsene und angesehene Leute gehören, so pfleget man daher Dinge, die man herunter und lächerlich machen will, Schöne Raritäten, schöne Spielwercke zu nennen." In Füsslin, *Der Guckkasten*, 8–10. The name *Raritätenkasten*, or "rarities box," prevailed in the first half of the eighteenth century, *Guckkasten* becoming more common in the latter half.

8. Such a "mondo nuovo" is seen in the carnivalesque opening scene of Ettore Scola's *Il mondo nuovo*, a period film set during the French Revolution. *Il mondo nuovo/La Nuit de Varennes*, dir. Ettore Scola (Triumph Releasing Corporation, 1982).

9. The Jesuit Zacharias Traber described such a peepshow in a treatise on optics, *Nervus opticus sive tractatus theoricus* (Vienna, 1675): the viewer looked through a small opening at a mirror, which was positioned to reflect figures attached to a wheel that rotated the figures into view.

10. Erkki Huhtamo, *Illusions in Motion: Media Archaeology of the Moving Panorama and Related Spectacles* (Cambridge, MA: MIT Press, 2013), 35–36.

11. According to John Zucchi, the chimney sweep and itinerant entertainer were misleading Savoyard stereotypes, both because Savoyard migrants performed many other jobs and because these jobs were performed by migrants from other parts of Italy; John E. Zucchi, *Little Slaves of the Harp: Italian Child Street Musicians in Nineteenth-Century Paris, London, and New York* (Buffalo: McGill-Queen's University Press, 1992), 42–44.

12. On street-crier prints, see Karen Beall, *Kaufrufe und Strassenhändler: Eine Bibliographie/ Cries and Itinerant Trades: A Bibliography* (Hamburg: Hauswedell, 1975). On street cries in music, see Maria Rika Maniates and Richard Freedman, "Street Cries," *Grove Music Online. Oxford Music Online*. Oxford University Press, accessed July 29, 2014, http://www.oxfordmusiconline.com/subscriber/article/grove/music/26931; and Jean-Rémy Julien, *Musique et publicité: Du Cri de Paris . . . aux messages publicitares radiophoniques et télévisés* (Paris: Flammarion, 1989).

13. For a survey of poetic texts written in the style of a Savoyard performance, see A. Kopp, "Schöne Spielewerk, schöne Rarität!" *Archiv für Kulturgeschichte*, vol. 2, ed. Georg Steinhausen (Berlin: Alexander Duncker Verlag, 1904), 296–317. "In Savoyard-tone" [in Savojarden-Tone] appears as a performance direction for the peepshowman in Joseph Richter, *Das schöne Milchmädchen, oder Der Guckkasten* (Vienna: Mit von Kurtzbekischen Schriften, 1796), 19.

14. Reinhard Keiser and C. H. Weidemann, *Le Bon Vivant, oder Die Leipziger Messe* (Hamburg: Gennagel, 1710). The opera was performed at the Theater am Gänsemarkt in Hamburg in 1710, but the music by Keiser is lost.

15. Gotthold Ephraim Lessing, "Fortsetzung des ein und Vierzigsten Briefes" (1758–1759), in *Werke und Briefe in zwölf Bänden*, ed. Gunter E. Grimm (Frankfurt am Main: Deutscher Klassiker Verlag, 1997).

16. Johann Christoph Kaffka, *Der Guk Kasten, oder das beste kommt zulezt* (Breslau, 1782), 97.

17. ". . . wenn das gekünstelte ha-ha-ha, he-he-he der Sänger von der Bühne seinen Abschied bekommen. . . . und da sie zugleich sehr wenige von den Ha-he-isten entlehnten geschwänzte Schneller, die man den neuen Geschmack nennet . . ." Georg Philipp Telemann, *24 Oden* (Hamburg: Christian Herold, 1741), 3; partially trans. in Ann Le Bar, "The Domestication of Vocal Music in Enlightenment Hamburg," *Journal of Musicological Research* 19 (2000): 100.

18. "Scozzonatore cioè menino de' forastieri per una città grande, e che loro và mostrando le rarità." Jacob and Wilhelm Grimm, *Deutsches Wörterbuch*, 32 vols. (Leipzig: S. Hirzel, 1971), 30:2286, http://www.woerterbuchnetz.de/DWB?bookref=30,2286,45.

19. Meredith Ellis Little, "Forlana," *Grove Music Online. Oxford Music Online*. Oxford University Press, accessed July 18, 2014, http://www.oxfordmusiconline.com/subscriber/article/grove/music/09980.

20. Raymond Monelle, *The Musical Topic: Hunt, Military, and Pastoral* (Bloomington: Indiana University Press, 2006), 215–217.

21. For example, Kotzebue's comedy *Pagenstreiche*, published in 1804 and first performed in 1810, included two characters who disguised themselves as Savoyard showmen, one carrying a peepbox, the other a barrel organ. August von Kotzebue, *Neue Schauspiele*, vol. 11 (Leipzig: Paul Gotthelf Kummer, 1804), 123. In 1808, E. T. A. Hoffmann composed music for a ballet titled "Arlequin," which contains a scene in which Arlequin exhibits a peepbox.

22. Christian Felix Weisse, *Christian Felix Weissens Selbstbiographie, herausgegeben von dessen Sohne Christian Ernst Weisse und dessen Schwiegersohne Samuel Gottlob Frische, mit Zusätzen von dem Letztern* (Leipzig: Georg Voss, 1806), 103; trans. Estelle Joubert, "Songs to Shape a German Nation: Hiller's Comic Operas and the Public Sphere," *Eighteenth-Century Music* 3 (2006): 226.

23. "Es ist eine Farce, bey der ich alles mögliche Lob erhalten habe, wenn man recht von ganzem Herzen lacht." Johann Benjamin Michaelis, *Operetten*, vol. 1 (Leipzig: Dyckischen Buchhandlung, 1772), [ii].

24. ". . . so wären solche kleine Stücke von einem Alte noch das beste Mittel, die Liebhaberey der Menge mit dem Geschmacke des ernsten Kenners zu vereinigen." Michaelis, *Operetten*, [i].

25. Bauman documents a number of competing premiere dates and locations for Neefe's *Amors Guckkasten*, including Leipzig, May 10, 1772, and Königsberg, February or August 1772. Thomas Bauman, *North German Opera in the Age of Goethe* (Cambridge: Cambridge University Press, 1985), 356.

26. Thomas Christensen, "Public Music in Private Spaces: Piano-Vocal Scores and the Domestication of Opera," in *Music and the Cultures of Print*, ed. Kate van Orden (New York: Garland Press, 2000), 68–69.

27. Breitkopf und Härtel, "Nachricht für Theaterdirektionen," *AmZ* 1/11 (April 1799), col. 57.

28. Bauman, *North German Opera*, 70.

29. Johann Adam Hiller, "Abhandlung über die Nachahmung der Natur in der Musik," in F. W. Marpurg, *Historisch-kritische Beyträge zur Aufnahme der Musik*, vol. 1 (Berlin, 1754–5/*R*), 534; trans. in Edward A. Lippman, *A History of Western Musical Aesthetics* (Lincoln: University of Nebraska Press, 1992), 118. Lippmann counts Hiller among progressive thinkers who emphasized the sensuous and expressive dimensions of music, in contrast to Krause, Marpurg, and Sulzer, who propounded theories of the affections as objective emotional states to be depicted and aroused by music.

30. Charles Batteux, "Les beaux-arts réduits à un même principe," in *Music and Aesthetics in the Eighteenth and Early-Nineteenth Centuries*, ed. Peter le Huray and James Day (New York: Cambridge University Press, 1981), 40–42, 49.

31. Jean-Jacques Rousseau, "Dictionnaire de musique," in *Music and Aesthetics in the Eighteenth and Early-Nineteenth Centuries*, 86–87.

32. "Also in Attilio Regeol [Regolo], in O qual siamme di gloria etc. tobt ein solcher Aufruhr, den ein Herz, durch Begierde nach Ruhme und Ehre erhitzt, in allen Adern zu fühlen nur im Stande ist." Hiller, "Abhandlung über die Nachahmung der Natur in der Musik," 535.

33. Trans. in Lippman, *A History of Western Musical Aesthetics*, 119.

34. Ibid.

35. Christian Gottlob Neefe, *Amors Guckkasten: Eine komische Operette*, ed. Gerhart von Westerman (München: Drei Masken Verlag, 1922).

36. See, for example, E. T. A. Hoffmann's imitation of a hurdy-gurdy for the scene in which Arlequin exhibits a peepbox in his *Arlequin Ballettmusik* (1808).

37. On the status of "rarities" and "curiosities," see Neil Kenny, *Curiosity in Early Modern Europe: Word Histories* (Wiesbaden: Harrassowitz, 1998), 14; Lorraine Daston and Katherine Park, *Wonders and the Order of Nature, 1150–1750* (New York: Zone Books, 1998).

38. "Hier schlug er seine Bude auf, und rief den Pöbel zusammen, seine schöne Raritäten zu schauen. Das Volk drang hinzu, Mägde und Knechte, Jungfrauen und Wittwen, Kinder und Greise. Der Teufel gaukelte ihnen allerley histörchen vor, die er mit frommen Erläuterungen und moralischen Sprüchen begleitete." The phrase "schöne Raritäten zu schauen" and list of diverse viewers are typical pieces of peepshowman rhetoric. Friedrich Maximillian Klinger, *Fausts Leben,*

Thaten und Höllenfahrt (St. Petersburg: Johann Friedrich Kriele, 1791), 224; the translation is taken from Friedrich Maximillian Klinger, *Faustus: His Life, Death, and Descent into Hell*, trans. George Borrow (London: W. Simpkin and R. Marshall, 1825), 142.

39. "Die englische Angelika sah aus dem Fenster, und da sie den Teufel mit einem so frommen Tone, die Vorspieglung seiner histörchen ableyern hörte, fühlte sie ein unwiderstehliche Versuchung, die Wunder des Kasten zu sehen, und dem frommen Greise ein Almosen zufließen zu lassen." Ibid., 224–225; trans. in ibid., 142–143.

40. ". . . der Teufel leyerte seine Alltagssprüche herunter, und gaukelte ihr stufenweis die Scenen der Liebe, bis zu den ausschweifendsten Vorspiegelungen der Wollust und des sinnlichen Genussses, vor. Führte ihre Fantasie so rasch und unmerklich vom Geistigen zum Sinnlichen hinüber, daß sie die Schattirung kaum gewahr werden konnte." Ibid., 225; trans. ibid., 143.

41. "Nun deckte sie mit beiden Händen ihre Augen, floh nach ihrem Schlafzimmer, und sank Fausten in die Arme." Ibid., 227; trans in ibid., 144.

42. "Da sahen sie einen wunderlichen Mann hinter einem seltsamen Kästlein stehen; das sahe wie ein kleines Häuslein voller runder Fenster aus, mit bunten Vorhängen da und dort überkleidet." Friedrich Baron de la Motte Fouqué, "Die Kuckkasten," in *Kinder-Mährchen*, by E. W. Contessa, Friedrich Baron de la Motte Fouqué, and E. T. A. Hoffmann (Berlin: G. Reimer, 1839), 254.

43. Ibid., 255.

44. "Dieser lockenden Versprechungen ungeachtet, schien sich eben Riemand herbeimachen zu wollen, und auch Karl Grünbaum wäre wohl mit seinen Gefährten ohne weiteres vorübergerannt, wenn nicht der Fremde eine kleine Drehorgel zu spielen anhob, um welche sich viele helle Silberglöcklein im besten Takte herdrehten, und die Weise der obigen Verslein recht zierlich wiederholten." Ibid.

45. Karol Berger, *Bach's Cycle, Mozart's Arrow* (Berkeley: University of California Press, 2007), 1.

46. The technological confusion here is rooted in a terminological one: as Laurent Mannoni clarifies, "the magic lantern and the peepshow box (known in France as the *boîte à vues d'optique*; in Germany as *Bilder-Guckkasten*; and in Italy as the *camera otticha* or *Il mondo nuovo*) were two quite distinct travelling 'curiosities.' The iconography of the eighteenth and nineteenth centuries cheerfully mixed the two devices under the generic term 'magic lantern,' although they were quite different," the magic lantern—as we shall see in chapters 4 and 5—being an image projection rather than a peep-view apparatus. Laurent Mannoni, *The Great Art of Light and Shadow: Archaeology of the Cinema*, trans. Richard Crangle (Exeter: University of Exeter Press, 2001), 86. On the technology referenced in Goethe's *Faust II*, see Neil M. Flax, "Goethe's Faust II and the Experimental Theater of His Time," *Comparative Literature* 31 (1979): 154–166.

47. Erkki Huhtamo, "The Pleasures of the Peephole: An Archaeological Exploration of Peep Media," *Book of Imaginary Media*, ed. Erick Kluitenberg (Rotterdam: De Balie, 2006), 100.

48. "Niedergedrückt von den Gräueln des Krieges, vertraut geworden mit allem Elende, suchte man nur Erheiterung in den gröblichst aufreizenden Kunstlüsten. Das Theater war zum Guckkasten, in welchem man gemächlich—die schön-beglückende Gemüths-Unruhe beym wahren Genusse eines Kunstwerks ängstlich vermeidend—eine Scenen-Reihe vor sich abhaspeln liess; zufrieden, durch triviale Spässe und Melodien gekitzelt worden zu seyn, oder geblendet durch Maschinen-Unfug ohne Zweck und Sinn." Carl Maria von Weber, "Ueber die Oper, Undine," *Allgemine musikalische Zeitung* 19/12 (March 19, 1817): col. 204; trans. (here modified) Raptus Association for Music Appreciation, "Hoffmann-Undine," http://www.raptusassociation.org/hoffm_undinewebere.html. This translation takes fewer liberties than that found in Carl Maria von Weber, *Carl Maria von Weber: Writings on Music*, ed. John Warrack, trans. Martin Cooper (Cambridge: Cambridge University Press, 1981), 202.

49. "Ein in sich abgeschlossenes Kunstwerk, wo alle Theile und Beyträge der vewandten und benutzten Künste ineinanderschmelzend verschwinden, und auf gewisse Weise untergehend—eine neue Welt bilden." Weber, "Ueber die Opera, Undine," col. 203; trans. ibid.

50. "Uebelgesinnte wollen den Decorationen viel zuschreiben; wenn aber Ref. bemerkt, dass in andern Stücken, wo dieses der Fall ist, die Leute nur diese Momente ablauern und dann wieder

gehen, hier aber mit steter, gleicher Aufmerksamkeit von Anfang bis zu Ende bleiben: so beweist dies schon hinlänglich für das Interesse, das ihnen die Sache selbst einflösst." Weber, "Ueber die Opera, Undine," col. 207; trans. ibid.

51. Martin Nedbal, "Live Marionettes and Divas on the Strings: *Die Zauberflöte's* Interactions with Puppet Theater," *Opera Quarterly* 28 (2012): 21.

52. Anthony Newcomb, "New Lights on Weber's Wolf Glen," in *Opera and the Enlightenment*, ed. Thomas Bauman and Marita McClymonds (Cambridge: Cambridge University Press, 1995), 76.

53. Crary, "Géricault, the Panorama, and Sites of Reality," 8–9.

54. Theodor Adorno, "Bourgeois Opera," in *Sound Figures*, trans. Rodney Livingstone (Stanford, CA: Stanford University Press, 1999), 21; Carolyn Abbate, *In Search of Opera* (Princeton, NJ: Princeton University Press, 2001), 6; Grover-Friedlander, *Vocal Apparitions*, 7.

55. "Non mai volga ver lei gli avidi lumi,/Che di perdia eterna/Gli sia certa cagion un solo sguardo." Alessandro Striggio, *Orfeo: Favola in Musica*, trans. Gilbert Blin (Boston Early Music Festival, 2012), http://www.bemf.org/media/1213_opera_libretto.pdf.

56. Letter to Goethe (October 27, 1809), emphasis added. Trans. in Bonds, *Music as Thought*, 24.

57. Ibid., 24.

58. See, for example, Bonds's glosses: "One does not understand the Absolute; the best one can hope for is a glimpse of it" (ibid., 58); "Like earlier writers, Hoffmann perceived music as occupying an altogether separate sphere beyond the phenomenal, thereby endowing musical works with the power to provide a glimpse of the infinite" (ibid., 27).

59. Bonds does not address the cover image within *Music as Thought*, but the same image is reproduced in his music appreciation text *Listen to This*, where it is said to capture "the Enlightenment's ideal of the mind—the seat of reason—as the focal point of all perception." Mark Evan Bonds, *Listen to This* (Upper Saddle River, NJ: Pearson Prentice Hall, 2009), 171.

CHAPTER THREE

1. "Bürgers Leonore von André in Musik gesezt und in Chinesischem Schattenspiel aufgeführt von der Fräulein von B*** zu Regensburg 1781," *Litteratur- und Theater-Zeitung* 37 (September 15, 1781): 577.

2. Friedrich Kittler, *Discourse Networks 1800/1900*, trans. Michael Metteer (Stanford, CA: Stanford University Press, 1990), 109.

3. Celeste Langan, "Understanding Media in 1805: Audiovisual Hallucination in 'The Lay of the Last Minstrel,'" *Studies in Romanticism* 40 (2001): 49–70. Martha Woodmansee discusses the impact Bürger's "Lenore" had on Scott in *The Author, Art, and the Market: Rereading the History of Aesthetics* (New York: Columbia University Press, 1994).

4. See notes 85 and 93 to this chapter for the sources of these quotes.

5. Richard Daniel Altick, *The Shows of London* (Cambridge, MA: Belknap Press, 1978), 117–119; Gerd Eversberg, "'*Ombres chinoises*': Zur Geschichte eines Medienspektakels seit dem siebzehnten Jahrhundert," in *Die Mobilisierung des Sehens*, ed. Harro Segeberg (Munich: Wilhelm Fink Verlag, 1996), 45–68.

6. "Je me souviens d'avoir été singulièrement émerveillé dans mon enfance par le noble jeu appelé schattenspiel en allemand, représenté par des comédiens ambulans avec beaucoup de succès." Friedrich Melchior Grimm, *Correspondance littérarie, philosophique et critique de Grimm et de Diderot, depuis 1753 jusqu'en 1790*, vol. 7 (Paris: Furne, 1812), 49. The publication prompting Grimm's remarks was Pierre Marion de Salins, *L'heureuse pêche: comédie pour les ombres a scenes changeantes: représentée pour la premiere fois au Château de ***, le 22 décembre 1767* (Paris: Chez Le Jay, 1770).

7. "Après l'Opéra français je ne connais pas de spectacle plus intéressant pour les enfans; il se prête aux enchantemens, au merveilleux et aux catastrophes les plus terribles." Grimm, *Correspondance littérarie*, 49.

8. On Grimm's satire of French opera, see Heartz, *Music in European Capitals*, 710–711.

9. Fan Pen Li Chen, *Chinese Shadow Theatre: History, Popular Religion, and Women Warriors* (Montreal: McGill-Queen's University Press, 2007).

10. Dorinda Outram, *Panorama of the Enlightenment* (Los Angeles: J. Paul Getty Museum, 2006), 155–161.

11. François Dominique Séraphin called himself "author and inventor of the ombres chinoises" in advertisements; see *Feu Séraphin: Histoire de ce spectale depuis son origine jusqu'a sa disparition, 1776–1870* (Lyon: N. Scheuring, 1875), 7. For further information on François Dominique Séraphin and *ombres chinoises*, see Denis Bordat and Francis Boucrot, *Les théâtres d'ombres: histoire et techniques* (Paris: L'Arche, 1956).

12. Altick, *Shows of London*, 118.

13. ". . . hier noch niemals gesehen/LES OMBRES CHINOISES." Broadside reproduced in Hans Purschke, "Puppenspiel und verwandte Künste in der Reichsstadt Nürnberg," *Mitteilungen des Vereins für Geschichte der Stadt Nürnberg* 68 (1981): 241.

14. "Les canards l'ont bien passée, Tire lire, lire." Trans in George Speaight, *The History of the English Puppet Theater* (Carbondale: Southern Illinois University Press, 1990), 146. The text has been credited to the French playwright Charles-Jacob Guillemain, the scenario being derived from a popular fable; see *Feu Séraphin*, 2, 33–36, and Guillemain, *Le Pont Cassé*, ed. Denis Bordat and Francis Boucrot (Paris: Editions du Scarabée, 1954).

15. Philip Astley, *The favourite airs for the violin, German flute, harp, or harpsichord, with a bass; together with the scene of the broken-bridge in the Ombres chinoises* (London, 1780). Astley, best known for his equestrian feats and circus, also exhibited *ombres chinoises* from about 1776 to 1789, which he mixed with other entertainments advertised for "the Nobility, Gentry, and Others." See Austin Dobson, "Ombres Chinoises," *The Cornhill Magazine* 11 (1901): 758.

16. Max von Boehn, *Puppets and Automata*, trans. Josephine Nicoll (New York: Dover, 1972), 116.

17. "Ce spectacle, où règne la gaîté est toujours caractérisé par la décence." Quoted in Bordat and Boucrot, *Les théâtres d'ombres: histoire et techniques*, 78.

18. "Wird hiedurch bekannt gemacht, daß Herr Marquis, ein Italianer, und durch ganz Europa berühmter Mechanikus, dahier angekommen sey, und die Ehre hat, heute abermal in hiesiger Stadt sein berühmtes Kunstkabinet sehen zu lassen, betitelt: LES GRANDES OMBRES CHINOISES." Ad, dated December 17, 1781, reproduced in Purschke, "Puppenspiel und verwandte Künste in der Reichsstadt Nürnberg," plate 7.

19. Trans. in Philip P. Wiener, "Leibniz's Project of a Public Exhibition of Scientific Inventions," *Journal of the History of Ideas*, 1 (1940): 232–240.

20. Ibid., 234.

21. Ibid., 239–240.

22. Ibid., 239.

23. [Edmé-Gilles] Guyot, *Nouvelles récréations physiques et mathématiques*, vol. 3, part 6 (Paris: Gueffier, 1774), 174–176; [Edmé-Gilles] Guyot, *Neue physikalische und mathematische belustigungen, oder Sammlung von neuen kunststücken, zum vergnügen*, vol. 6, trans. John Christoph Thenn (Augsburdge: Eberhard Kletts, 1776), 162–165.

24. Johann Christian Wiegleb, *Die natürliche Magie: aus allerhand belustigenden und nützlichen Kunststücken bestehend*, vol. 3 (Berlin and Stettin: Friedrich Nicolai, 1779), 173. This book went on to numerous reprintings as the first volume of Wiegleb's multivolume compendium of popular science (read by E. T. A. Hoffmann, as we shall see in chapter 5). A similar description of the Chinese shadow-play also appeared in Johann Samuel Halle, *Magie, oder die Zauberkräfte der Natur, so auf den Nutzen und die Belustigung angewandt worden*, vol. 1 (Wien: Johann Thomas Edlen von Trattnern, 1787), 267–269.

25. "Il est essentiel en les faisant agir, de faire quelque dialogue qui suive exactement leurs gestes, & on doit même imiter le bruit (lorsqu'il est convenable), c'est-à dire, que si on fait tomber une figure à bas d'une échelle, il faut imiter le bruit qu'une échelle fait en tombant, &c." Guyot, *Nouvelles récréations physiques et mathématiques*, 176. "Es ist auch unumgänglich nöthig, wenn man sie sich bewegen lässet, ein Gespräch dabey zu unterhalten, das sich genau zu ihren Gebehrden und Bewegungen schickt, ja man muß auch selbst wenn es thunlich ist, das Geräusche nachahmen, das ist, wenn man eine Figur von der Leiter herabfallen lässet, so muß man das Geräusche

nachahmen, welches der Fall einer Leiter machet." Wiegleb, *Die natürliche Magie*, 175. The choice of falling off a ladder for illustrative purposes here likely owes to one of Séraphin's vignettes, "Le Bucheron," in which a man falls from a ladder. See [L.-V. Flamand-Grétry], *Théâtre de Séraphin ou des ombres chinoises*, vol. 2 (Paris: Desmarest, 1809), 1–17.

26. The journal spells out "Regensburg" where the pamphlet reads "R***." Otherwise, the textual content of the two sources is identical.

27. "Herrn Bürgers leibliches Kind und, Herrn Andres Pfegotochter, Leonore, hat zu Regensburg die lezte Erziehung erhalten, da Sie von der Fräulein von B. im Chinesischen Schattenspiel, nach Ihrer eigenen Erfindung und Ausarbeitung mit Musik gegeben worden ist." "Bürgers Leonore," *Litteratur- und Theater-Zeitung*, 577.

28. On traditional approaches to lied criticism and performance, see Jennifer Ronyak, "Introduction," Colloquy: "Studying the Lied: Hermeneutic Traditions and the Challenge of Performance," *Journal of the American Musicological Society* 67 (2014): 543–549.

29. For this and all subsequent quotations from the ballad, see Gottfried August Bürger, *Bürgers Werke in Einem Band*, ed. Lore Kaim-Kloock and Siegfried Streller (Weimar: Volksverlag, 1962), 64 (60–68); translations based on those in Gottfried August Bürger, *Lenore, Ballade von Bürger: in Drei Englischen Übersetzungen* (Göttingen: Johann Christian Dieterich, 1797) and the notes to *Liszt: The Recitations with Pianoforte* (Hyperion CDA67045, 1996), 7–19.

30. "Praeter propter können Sie hieraus den Ton errathen, welcher, wie ich mir schmeichele, in der Folge noch populärer und balladenmäßiger ist und seyn wird. Der Stoff ist aus einem alten Spinnstubenliede genommen." Letter to Boie (Gelliehausen, May 10, 1773), in *Bürger's Sammtliche Werke*, ed. August Wilhelm Bohtz (Göttingen: Dieterichschen, 1835), 463.

31. Trans. in Maike Oergel, *The Return of King Arthur and the Nibelungen: National Myth in Nineteenth-Century English and German Literature* (Berlin: Walter de Gruyter, 1998), 39.

32. Trans. in Renata Schellenberg, "The Impact of Ossian: Johann Gottfried Herder's Literary Legacy," in *The Voice of the People: Writing the European Folk Revival, 1760–1914*, ed. Matthew Campbell and Michael Perraudin (New York: Anthem Press, 2012), 13.

33. David E. Wellbery, *The Specular Moment: Goethe's Early Lyric and the Beginnings of Romanticism* (Palo Alto, CA: Stanford University Press, 1996), 189–191. Herder's ideas about the vivacity of folk culture in contrast to the deadness of print culture are also discussed in Schellenberg, "The Impact of Ossian," 9–20, and Isaiah Berlin, *Three Critics of the Enlightenment: Vico, Hamann, Herder*, ed. Henry Hardy (Princeton, NJ: Princeton University Press, 2013), 240–244.

34. William Alfred Little, *Gottfried August Bürger* (New York: Twayne Publishers, 1974), 104.

35. Woodmansee, *The Author, Art, and the Market*, 63.

36. Gotthold Ephraim Lessing, *Laocoon*, trans. Robert Phillimore (London: Macmillan, 1874), 160–161.

37. Johann Christoph Adelung, *Ueber den deutschen Styl* (Berlin: C. F. Voss, 1785), 307; trans. in Matthew Riley, *Musical Listening in the German Enlightenment: Attention, Wonder, and Astonishment* (Burlington, VT: Ashgate, 2004), 134.

38. Woodmansee, *The Author, Art, and the Market*, 61–63.

39. Quoted in Little, *Gottfried August Bürger*, 106.

40. Ibid.

41. "Er hielt nämlich, wie von ungefähr, eine Reitgerte in der Hand, und als er an die Stelle kam: 'Rasch auf ein eisern Gitterthor/Gings mit verhängtem Zügel;/Mit schwanker Gert ein Schlag davor/Zersprengte Schloß und Riegel,' schlug er damit an eine gegenüber stehende Thür. Stolber, damals ein Jüngling von entzündbarer Einbildungskraft, durch die vorhergehende Schilderung schon ganz ergriffen, sprang hiebei mit Entsetzen auf, als ob die geschilderte Sache wirklich unter seinem Augen vorginge." August Wilhelm von Schlegel, *Kritische Schriften*, vol. 2 (Berlin: G. Reimer, 1828), 40–41.

42. "Eine Geschichte, welche die getäuschten Hoffnungen und die vergebliche Empörung eines menschlichen Herzens, dann all Schauer eines verzweiflungsvollen Todes in wenigen leicht faßlichen Zügen und lebendig vorüberfliehenden Bildern entfaltet, ist ohne erkünsteltes Beiwerk, ohne vom Ziel schweisende Auschmückungen in die regste Handlung, und fast ganz in wechselnde

Reden gesezt, während welcher man die Gestalten, ohne den Beistand störender Schilderungen, sich bewegen und gebehrden sieht." Ibid., 39.

43. Woodmansee, *The Author, Art, and the Market*, 66.

44. "Es sollte meine gröste Belohnung seyn, wenn es recht balladen-mäsig und simpel componirt, und dann wieder in den Spinnstuben gesungen werden könnte. Ich wollte ich könnte die Melodie, die ich in der Seele habe, dem Componisten mit der Stimme angeben!" Letter to Boie (Gelliehausen, May 10, 1773) in *Bürger's Sammtliche Werke*, 463.

45. Christian Felix Weisse, *Komische Opern*, vol. 3 (Carlsruhe: C. G. Schmieder, 1778), 2v-3r; trans. in Joubert, "Songs to Shape a German Nation," 216.

46. Johann Abraham Peter Schulz, *Lieder im Volkston*, vol. 2 (Berlin: Georg Jacob Decker, 1785); trans. in Gramit, *Cultivating Music*, 67.

47. Friedrich Reichardt, "An junge Künstler," *Musikalisches Kunstmagazin* 1 (1782): 3; trans. in Gramit, *Cultivating Music*, 74.

48. On the relationship of volkslied discourse to the precarious condition of composers in an increasingly market-driven economy, see Gramit, *Cultivating Music*, 73–79.

49. Friedrich Wilhelm Weis, *Lieder mit Melodien II* (Lübeck, 1776); Georg Wilhelm Gruber, *Bürger's Gedichte für das Klavier und die Singstimme gesezt* (Nuremberg, 1780); Johann Kirnberger, *Gesänge am Clavier* (Berlin und Leipzig: George Jacob Decker, 1780); and Georg Friedrich Wolf, *Lieder mit Melodien* (Nordhausen, 1781). A setting by Ignaz von Beecke, mentioned by C. F. Schubart in 1775 as being "in true ballad tone," likely existed only in manuscript and is now lost. See Max Friedlaender, *Das deutsche Lied im 18. Jahrhundert: Quellen und Studien*, vol. 1 (Stuttgart and Berlin: J. G. Cotta'sche, 1902), 219.

50. ". . . in dem Wunsch einiger Herren Subscribenten, daß die meisten dieser Lieder mit Vor- und Nachspielen des Klaviers möchten versehen werden, weil sie schon einige davon auf solche Art von mir besatzen." Gruber, *Bürger's Gedichte*, 2.

51. Georg Wilhelm Wegner, *Schauplatz Vieler Ungereimten Meynungen und Erzehlungen*, vol. 1 (Berlin and Leipzig: Ambrosius Haude, 1735), 225; trans. in Tom Cheesman, *The Shocking Ballad Picture Show: German Popular Literature and Cultural History* (Oxford: Berg Publishers, 1994), 70.

52. Ibid., 19 and chapter 2.

53. Cheesman, *The Shocking Ballad Picture Show*, 18. Cheesman covers the history of the ballad picture show in Germany from its sixteenth-century beginnings to its waning days in the early twentieth century.

54. Ibid., 73.

55. "Am stärksten aber wird das Volk gerührt von allem, was unter seine Augen gebracht wird. Weit mehr als eine ausführliche Beschreibung zieht ein gesudeltes Gemählde, ein kindischer Holzschnitt den dunkeln Menschen an. Und wie viel Tausende sind, die in dem vortrefflichsten Bilde nur das Mährchen erblicken. Die großen Bilder der Bänkelsänger drücken sich weit tiefer ein als ihre Lieder, obgleich auch diese die Einbildungskraft mit starken Banden fesseln." The passage comes from an early version of *Wilhelm Meisters Lehrjahre*, dated to between 1776 and 1785 and first published as Johann Wolfgang von Goethe, *Wilhelm Meisters theatralische Sendung*, ed. Hary Mayne (Berlin: J. G Cotta'sche Buchhandlung Nachfolger, 1911), 109; partially trans. in Cheesman, *The Shocking Ballad Picture Show*, 16.

56. ". . . es kommt darauf an den Hörer in die Stimmung zu versetzen welche das Gedicht angibt, in der Einbildungskraft bilden sich alsdann die Gestalten nach Anlaß des Textes, sie weiß nicht wie sie darzu kommt." The context here is Goethe's praise for Zelter's subtle style of word-painting, as exemplified in his songs "Johanna Sebus," "Um Mitternacht," and "Über allen Gipfeln ist Ruh." Letter to Zelter (Carlsbad, May 2, 1820); *Briefwechsel zwischen Goethe und Zelter in den Jahren 1799 bis 1832*, vol. 20, no. 1, of Johann Wolfgang von Goethe, *Sämtliche Werke nach Epochen seines Schaffens* (Münchner Ausgabe), ed. Karl Richter et al. (Munich: Carl Hanser, 1991), 599; trans. Lorraine Byrne Bodley, *Goethe and Zelter: Musical Dialogues* (Burlington, VT: Ashgate, 2009), 266.

57. "Die Lenore ist ein so vortreffliches Stück der Dichtkunst, so tonvoll und musikalisch, daß michs wunder nahm, warum sie noch kein geschickter Meister in Musik brachte. Herr Haupt-

mann *Beeke*, dieser große praktische und theorestische Musiker hat eine Musik hinzugesetzt, im wahren Balladen Ton. Da aber alle Strophen einerley Melodie haben; so konnt' er alle Schönheiten des Originals nicht erschöpfen. Herr André ist also der erste, der die ganze Lenore Strophe für Strophe in Musik gesetzt hat, und zwar so, daß man wohl sieht, er könne dem Dichter nachempfinden; nur scheint er mir nicht musikalische Kunst genug zu haben, denn man findet da und dort sehr anstößige Stellen." Christian Friedrich Schubart, "Lenore, von Bürger, in Musik gesetzt von André," *Deutsche Chronik auf das Jahr 1775*, Stück 63 (August 7, 1775): 502.

58. "Seine melodien sind leicht, gefällig, und drücken sich von selbst dem Gedächtnise ein . . . Die Tochter spielt sie auf dem Flügel, und singt darzu; der Bruder nimmt die Flöte oder die Geige, und der Vater streicht dem Baß. Welch ein angenehmes familienconcert in den Winterabenden!" Christian Friedrich Schubart, "Auserlesene scherzhafte und zärtliche Lieder, in Musik gesetzt von Johann André," *Deutsche Chronik auf das Jahr 1775*, Stück 3 (January 9, 1775): 22.

59. "Der Gesang strömt freywillig aus einem gerührten Herzen, hat schon sein Beet, das ihm die Natur grub, und bracht keinen von den Marpurgs und Kirnbergern mit Hacken und Schaufeln mübsam gegrabenen Kanal. Wir haben noch Volkslieder, die über hundert Jahr alt sind; aber wie ungekünstelt, wie leicht sind sie auch! Ihr Erfinder scheint die Noten aus dem Herzen gestohlen zu haben." Ibid.

60. "Wie kommts, dacht ich offt, daß die von berlinischen Tonkünstlern so vortrefflich in Musik gesezten Lieder der Deutschen fast nirgends gespielt, gesungen, goutirt werden? Sie sind zu schwerfällig, und mit einer zu ängstlichen Gewissenhaftigkeit niedergeschrieben." Ibid. The Berlin composers Schubart had in mind likely included Agricola, C. P. E. Bach, J. G. and C. H. Graun, Kirnberger, Krause, Marpurg, and Quantz—but not Schulz and Reichardt, whose song collections postdate this review. The former group of composers contributed to the three-volume collection *Berlinische Oden und Lieder*, published between 1756 and 1763, and are now often dubbed part of the "first Berlin school." See Heartz, *Music in European Capitals*, 425.

61. Marjorie Hirsch, *Schubert's Dramatic Lieder* (Cambridge: Cambridge University Press, 1993), 68.

62. Hirsch discusses this structural feature of Schubert's dramatic ballads in ibid., 87.

63. On musical signifiers of the oracular and supernatural, see Buch, *Magic Flutes & Enchanted Forests*.

64. "Indessen drücken die ziehenden Noten in den Worten: *Gedult, Gedult, wenn's Herz auch bricht*, das Todtengeheul des Geistes sehr gut aus." Schubart, "Lenore," 503.

65. "Kennst du schon dein Lenore von André in Musik gesezt? . . . O Bürger! Bürger! wärft du doch da gewesen! Solche Herrlichkeit der Musik, solche Kraft des Gesangs! Wie jeder Gedanke ganz ergriffen ist, und ganz ausgedrückt! Voll Wahrheit! Voll Natur! Einige Stellen sind über allen Ausdruck vortreflich. Wie hats mein Herz gelabt!" Biester to Bürger (Lübeck, Winter 1776), in *Briefe von und an Gottfried August Bürger*, ed. Adolf Strodtmann (Berlin: Gebrüder Paetel, 1874), 386–387.

66. "Du glaubst nicht, was beide Gerstenberg für hohe musikalische Talente haben. Sie treibt das Klavierspielen mit eifriger Passion; und hat eine sanfte, schöne, melodiereiche, biegsame, ausdaurende Stimme . . . Sie spielte; und sie und Er sangen, denn die Lenore ist als Duett gesezt." Ibid. Inaccurate in its description of André's setting, Beisler's claim that it is a duet suggests he encountered it only as a listener, without viewing the score; and it attests to the dramatic-scenic implications of the setting, which performers were ready to realize through multiple voices.

67. Johann Wolfgang von Goethe, *The Autobiography of Goethe. Truth and Poetry: From My Own Life*, vol. 2, trans. [unnamed] (London: George Bell and Sons, 1882), 83.

68. On Goethe's legendary performance abilities, and preference for a restrained style of poetic recitation to highly theatrical declamation, see Jennifer Ronyak, "*Performing the Lied, Performing the Self: Singing Subjectivity in German, 1790–1832*" (PhD diss., Eastman School of Music, 2010), 160–162.

69. Goethe, *The Autobiography of Goethe*, 241.

70. "Ich habe mir aber von verschiedenen Musikern sagen lassen: die Komposition sei abscheulich." Bürger to Boie (November 14, 1775), in *Briefe von und an Gottfried August Bürger*, 258.

71. See Eric Schön, *Verlust der Sinnlichkeit, oder, Die Verwandlungen des Lesers: Mentalitätswandel um 1800* (Stuttgart: Klett-Cotta, 1987), 185–187; Jennifer Ronyak, "'Serious Play,' Performance, and the Lied: The Stägemann *Schöne Müllerin* Revisited," *19th-Century Music* 34 (2010): 141–167.

72. Deborah Hertz, "Public Leisure and the Rise of Salons," in *Early Modern Europe: Issues and Interpretations*, ed. James B. Collins and Karen L. Taylor (Malden, MA: Blackwell Publishing, 2006), 282–296.

73. "Bürgers Leonore," *Litteratur- und Theater-Zeitung*, 583.

74. ". . . nach Ihrer eigenen Erfindung und Ausarbeitung mit Musik gegeben worden ist." "Bürgers Leonore," *Litteratur- und Theater-Zeitung*, 577.

75. *Objects of Vertu and Miniatures: Christie's South Kensington, Tuesday, 7 July 1998 at 10:30 A.M.* (South Kensington: Christie's, 1998), 10.

76. The publication of André's fourth "Lenore" is dated to 1790 or 1791 by its plate number; see Dietrich Manicke (ed.), *Balladen von Gottfried August Bürger: In Musik gesetzt von André, Kunzen, Zumsteeg, Tomaschek und Reichardt* (Mainz: Verlag B. Schott's Söhne), 256. Following his debut tour as a child prodigy, Johann Nepomuk Hummel dedicated his op. 3, *Trois Airs avec Variations pour le Piano Forte ou Clavecin* (Vienna: Author, 1794), to "Madame le Baronne de Vrintz, née Baronne de Berberich."

77. "Das Orchester bestand aus einem Späthischen Flügel, und aus einem vortreslichen Englischen Forte und Piano. Beide waren in der zur Execution des Schattenspiels bestimmten Cammer angebracht. Auf lezterem wurde das Accompagnement von Andre, und auf ersterem der General Bass, nach dieser Composition, gespielt." "Bürgers Leonore," *Litteratur- und Theater-Zeitung*, 577–578.

78. "Den erzählenden Theil des Gedichts übernahm ein Bänkelsänger . . . Dieser fas im dunklen Zimmer, wo die Zuschauer versammlet waren, nächst der Rahme, auf welcher die Vorstellungen sich zeigten." Ibid., 578.

79. ". . . im Kostume von Hanns Sachs gekleidet." "Bürgers Leonore," *Litteratur- und Theater-Zeitung*, 578.

80. Dieter Borchmeyer, *Drama and the World of Richard Wagner* (Princeton, NJ: Princeton University Press, 2003), 186.

81. "Die Acten und Scenen des Stücks konnten sich nicht nach den Reglen der Kunst, sondern nach den Dekorationen richten, so die Romanze vorzeichnete." "Bürgers Leonore," *Litteratur- und Theater-Zeitung*, 578.

82. Ibid., 580.

83. On the onomatopoeic status of "hop hop hop" (as well as "trap trap trap" and "husch husch husch) for the galloping horse, see Andrew Piper, "Transitional Figures: Image, Translation, and the Ballad from Broadside to Photograph," in *Book Illustration in the Long Eighteenth Century: Reconfiguring the Visual Periphery of the Text*, ed. Christina Ionescu (Newcastle: Cambridge Scholars Publishing, 2011), 171.

84. "Du kannst daher diese Spiele auch in solche Bildungen treiben, welche sich niemals physisch, d.h. von einem Schattenspieler bewerkstellen ließe, wenn nur die Phantasie folgen kann. Optische Spiele für das innere Auge. Auch zieht mich das bei solchen Spielen an, daß sie sich nicht begnügen, die Mechanismen des Lebens nachzuahmen, sondern durch ihre Zauberei ganz neue Verhältnisse hervorbringen." Letter to Kerner (September 27, 1809), in Thomas Sebastian, "Schattenspiel und laterna magica als poetologische Metaphern—Antizipation der Traumfabrik bei Justinus Kerner und Achim von Arnim," *Michigan Germanic Studies* 23 (1997): 8.

85. The relevant lines of text are "On the right and on the left/passing in front of their eyes/how flew heaths, meadows and country" (Zur rechten und zur linken Hand,/Vorbei vor ihren Blicken/Wie flogen Anger, Heid und Land!); "How flew on the right, how flew on the left/Mountains, trees and hedges" (Wie flogen rechts, wie flogen links/Gebirge, Bäum und Hecken!/Wie flogen links und rechts und links/Die Dörfer, Städt und Flecken!) "How flew, what the moon shone upon/How it flew in the distance!/As overhead flew/The sky and stars!" (Wie flog, was rund der Mond beschien,/Wie flog es in die Ferne!/Wie flogen oben überhin/Der Himmel und die Sterne!)

86. "Zuerst eignete Kirnberger sich die Lenore zu; und, einsichtsvoll, streng consequent, aber

auch kalt und trocken, wie er war, sah er in ihr blos die Ballade—blos die Gattung und äussere Form . . . Es lässt sich wol behaupten, dass kein Mensch auf Erden vermoch that, sich, nach sechzig solchen Noten, durch zwey und dreyssig achtteilige Strophen hindurch zu würgen." Friedrich Rochlitz, "Recension: Lenore, Ballade von G. A. Bürger," *Allgemeine musikalische Zeitung*, 15/41 (October 13, 1813): col. 674–675.

87. "*André*, der Vater, den Begriff der Ballade"; "Musik den Einzelnheiten des Textes sehr nahe gebracht ward"; "sogenannten Durchcomponirens"; "eine Art Cantate, wo jede Strophe ihre eigene Musik erhielt." Rochlitz, "Recension: Lenore, Ballade von G. A. Bürger," col. 674.

88. "Phantasie und Empfindung bekamen den Platz allein: und welchen Platz sie einmal erobert haben, den, weiss man ja, geben sie schwerlich wieder heraus, sondern erweitern und schmücken ihn nur immer mehr . . ." Ibid., 676.

89. ". . . das immer weiter um sich Greifen der Musik gegen die Ansprüche des Gedichts . . ." Ibid., 674.

90. Hirsch, for example, writes that "the earliest musical settings of German ballad texts were strophic, resembling Volkslieder," then gives as evidence strophic settings of "Lenore," all of which postdate André's through-composed setting. Hirsch, *Schubert's Dramatic Lieder*, 66. Offering an alternative history of the ballad, less indebted to the narrative Rochlitz seems to have inaugurated, James Parsons traces the through-composed ballad's origins to A. B. V. Herbing's *Musikalischer Versuch in Fabeln und Erzählungen des Herrn Professor Gellerts* (1759). James Parsons, "The Eighteenth-Century Lied," in *Cambridge Companion to the Lied*, ed. James Parsons (Cambridge: Cambridge University Press, 2004), 55.

91. Trans. in Hirsch, *Schubert's Dramatic Lieder*, 67.

92. Trans. in David E. Wellbery, *Lessing's Laocoon: Semiotics and Aesthetics in the Age of Reason* (Cambridge: Cambridge University Press, 1984), 117.

93. "Bürgers Leonore," *Litteratur- und Theater-Zeitung*, 583.

94. Otto Deutsch, *Schubert: A Documentary Biography*, trans. Eric Blom (New York: Da Capo Press, 1977), 906.

CHAPTER FOUR

1. Quoted in Bruce C. MacIntyre, *Haydn, The Creation* (New York: Schirmer Books, 1998), 257.

2. *Zeitung für die elegante Welt* (December 31, 1801); trans. in Landon, *Haydn: Chronicle and Works*, 5:186.

3. *Zeitung für die elegante Welt* (December 22, 1801); trans. in Landon, *Haydn: Chronicle and Works*, 4:600.

4. Quoted in Nicholas Temperley, *Haydn: The Creation* (New York: Cambridge University Press, 1991), 43.

5. Temperley, *Haydn*, 42.

6. Classic eighteenth-century texts on tone-painting include Johann Georg Sulzer, "Tone Painting," in *Aesthetics and the Art of Musical Composition in the German Enlightenment: Selected Writings of Johann Georg Sulzer and Heinrich Christoph Koch*, ed. Nancy Kovaleff Baker and Thomas Christensen (Cambridge: Cambridge University Press, 1995), 91; Johann Jakob Engel, "On Painting in Music," in *Strunk's Source Readings in Music History: The Late Eighteenth Century*, ed. Wye Jamison Allanbrook (New York: W. W. Norton, 1998), 229. For further discussion of the aesthetic discourse around tone-painting, see Will, *The Characteristic Symphony in the Age of Beethoven*, 129–155.

7. Will, *Characteristic Symphony*, 130.

8. Gillen D'Arcy Wood, *The Shock of the Real: Romanticism and Visual Culture, 1760–1860* (New York: Palgrave, 2001), 7.

9. The shift from individually hand-painted to mass-reproduced magic-lantern slides began around 1823, when English optician Philip Carpenter deployed a method of transfer-printing from copper plates. Stephen Herbert, "Magic Lantern Shows," in *Encyclopedia of Early Cinema*, ed. Richard Abel (New York: Routledge, 2005), 406.

10. Carl Friedrich Zelter, "Die Schöpfung: Ein Oratorium," *Allgemeine musikalische Zeitung* 4

(March 10, 1802): col. 389, trans. in Landon, *Haydn: Chronicle and Works*, 4:595. Johann Karl Friedrich Triest, "Bermerkungen über der Ausbildung der Tonkunst in Deutschland im achtzehnten Jahrhundert," *Allgemeine musikalische Zeitung* 3 (March 11, 1801): col. 409, trans. by Susan Gillespie as "Remarks on the Development of the Art of Music in Germany in the Eighteenth Century," in *Haydn and His World*, ed. Elaine Sisman (Princeton, NJ: Princeton University Press, 1997), 374.

11. "Magie heißt überhaupt die Kunst, Erfolge hervorzubringen, welche die natürlichen Kräfte der Körper zu übertreffen scheinen. Man theilte sonst diese Kunst in die natürliche Magie, bey welcher die wunderbar scheinenden Erfolge sich dennoch aus den Kräften und Gesetzen der Körper erklären ließen, und in die übernatürliche, welche die Mitwirkung der Geister erfordern sollte." Johann Samuel Traugott Gehler, *Physikalisches Wörterbuch oder Versuch einer Erklärung der vornehmsten Begriffe und Kunstwörter der Naturlehre*, vol. 3 (Leipzig: Schwickert, 1790), 89.

12. See Thomas L. Hankins and Robert J. Silverman, "The Magic Lantern and the Art of Demonstration," in *Instruments and the Imagination*, 43–71; Laura Mulvey, "Uncertainty: Natural Magic and the Art of Deception," in *Death 24x a Second: Stillness and the Moving Image* (London: Reaktion Books, 2006), 33–53. Also instructive here is Richard Altick's discussion of eighteenth-century "exhibitions of mechanical ingenuity," and the question "which was the spectator to admire more, the illusive representation of life or the sheer gadgetry involved in producing that illusion?" Altick, *Shows of London*, 64.

13. On the early history of the magic lantern, see Mannoni, *Great Art of Light and Shadow*, 28–73; Deac Rossell, *Laterna magica/Magic Lantern*, vol. 1 (Stuttgart: Füsslin Verlag, 2008), 1–63; Hankins and Silverman, *Instruments and the Imagination*, 43–49. Quote from Mannoni, *Great Art of Light and Shadow*, 54–55, citing William Moyneux, *Dioptrica Nova, a Treatise of Dioptricks*, 2nd ed. (London, 1709), 183–184.

14. Mme de Graffigny, letter to M. Devaux, December 11, 1738, in *Choix de Lettres du XVIIIe Siècle* (Paris, 1946), trans. Mannoni, *Great Art of Light and Shadow*, 108.

15. While images of indoor magic-lantern shows are many, narrative accounts are few. Tobias George Smollett's novel *The Adventures of Ferdinand Count Fathom* thus provides a rare illustration of how the multiple parts of the magic-lantern show unfolded in time. After the title character, disguised as a Savoyard magic lanternist and accompanied by a hurdy-gurdy player, proclaims his show outside a castle's gates, he is "invited into the courtyard, where the servants formed a ring, and danced to the efforts of his companion's skill; then he was conducted into the buttery, where he exhibited his figures on the wall and his princess on the floor." Tobias George Smollett, *The Adventures of Ferdinand Count Fathom*, 2 vols. (London, 1753), 2:325.

16. Note that the final chord of the minuet is misspelled (presumably a root-position C-major chord was meant). François Bedos de Celles, *The Organ-Builder*, trans. Charles Ferguson (Raleigh, NC: Sunbury Press, 1977), 308.

17. W. J. G. Ord-Hume, *Automatic Organs: A Guide to Orchestrions, Barrel Organs, Fairground, Dancehall & Street Organs including Organettes* (Atglen, PA: Schiffer Publishing Ltd., 2007), 225–243.

18. E. T. A. Hoffmann, "Automata," in *The Best Tales of Hoffmann*, ed. E. F. Bleiler, trans. Alexander Ewing (New York: Dover Publications, 1967), 96.

19. G. W. F. Hegel, *Aesthetics: Lectures on Fine Art*, trans. T. M. Knox (Oxford: Oxford University Press, 1975), 956.

20. Adelheid Voskuhl, *Androids in the Enlightenment* (Chicago: University of Chicago Press, 2013).

21. Terrance Riley, "Composing for the Machine," *European Romantic Review* 20 (2009): 376, 367.

22. Dolan, *The Orchestral Revolution*, 203.

23. See Annette Richards, "Automatic Genius: Mozart and the Mechanical Sublime," *Music & Letters* 80 (1999): 366–389. Mozart distinguished between musical machines on the basis of their sonic capacities, a factor that would complement more socially based distinctions: of a commission from a clockmaker (that distaste and boredom were preventing him from finishing); he wrote, "of course, if it were for a large clock and the thing sounded like an organ, then I might enjoy it;

but as it is, the instrument consists solely of shrill little pipes, which sound high-pitched and, to me, childish." Letter to Constanze (October 3, 1790), trans. Richards, "Automatic Genius," 367.

24. Johann Gottfried Mittag, *Historische Abhandlung von der Orgeln* (Lüneburg, 1756), trans. Deac Rossell in *Laterna magica*, 113.

25. Filippo Bonanni, *Gabinetto armonico pieno d'istromenti sonori indicati e spiegati* (Rome: Giorgio Placho, 1722). The Italian Bonanni identified barrel-organ-playing beggars as coming from Northern Europe, while elsewhere they were usually identified as Savoyard or Italian. Everywhere, the barrel organist was an outsider.

26. "Wilhelm, was ist unserem Herzen die Welt ohne Liebe! Was eine Zauberlaterne ist ohne Licht! Kaum bringst du das Lämpchen hinein, so scheinen dir die buntesten Bilder an deine weiße Wand! Und wenn's nichts wäre als das, als vorübergehende Phantome, so macht's doch immer unser Glück, wenn wir wie frische Jungen davorstehen und uns über die Wundererscheinungen entzücken." Johann Wolfgang von Goethe, *Die Leiden des jungen Werther* (Berlin: Neues Lebens, 1964), 69. Cited in Castle, *The Female Thermometer: Eighteenth-Century Culture and the Invention of the Uncanny* (New York: Oxford University Press, 1995), 156 (translation altered).

27. John H. Zammito, *The Genesis of Kant's Critique of Judgment* (Chicago: University of Chicago Press, 1992), 247. Andriopoulos has examined Kant's use of magic-lantern and related optical imagery throughout his philosophical writings, arguing that these played a "constitutive, rather than illustrative function for Kant's theory of transcendental illusion." Andriopoulos, *Ghostly Apparitions*, 46.

28. Johann Gottlieb Radlof, *Muustersaal aller teutschen Mund-arten, enthaltend Gedichte, prosaische Aufsätze und kleine Lustspiele in den verschiedenen Mund-arten aufgesetzt*, vol. 2 (Bonn: Büschler, 1822), 369–370. This treatise likely took the Schattenspiel Lied from an earlier publication, an exemplar for which appears in a collection of travelling fairground prints held by the Staatsbibliothek zu Berlin, Shelfmark Yd 7905 R.

29. Ibid.

30. "Dudeldei," *Deutsches Wörterbuch*, vol. 2, ed. Jacob Grimm, Wilhelm Grimm, et al. (Berlin: Hirzel, 1860), col. 1497; "Der (or das) Dudeldei," *An Historical Dictionary of German Figurative Usage*, 6 vols., ed. Keith Spalding (Oxford: B. Blackwell, 1952), 2:508.

31. "Die Orgelei," *An Historical Dictionary of German Figurative Usage*, 4:1817.

32. ". . . alten savoyarden-mäßige Orgelum orgelei, wovon Worttext und Melodie gleich schlecht sind." Gottfried August Bürger, "Ankündigung (einer Neubearbeitung von 'Tausend und eine Nacht')," in *Sämtliche Werke*, ed. Günter und Hiltrud Häntzschel (Munich: Carl Hanser Verlag, 1987), 737.

33. "Doktor: Laßt ihn rein kommen; Amtmann: Bringt den Schirm heraus; Doktor: Thut die Lichter aus/Sind ja in einem honetten Haus." Johann Wolfgang von Goethe, *Jahrmarktsfest zu Plundersweilern*, ed. Max Herrmann (Berlin: Weidmannsche Buchhandlung, 1900), 263.

34. Ibid., 264; trans. based on Eric Hadley Denton, "The Technological Eye: Theater Lighting and *Guckkasten* in Michaelis and Goethe," in *The Enlightened Eye: Goethe and Visual Culture*, ed. Evelyn K. Moore and Patricia Anne Simpson (New York: Rodopi, 2007), 256–257.

35. "Dank sey dir, lieber Herre Gott/Orgelum, orgeley,/Dudeldumdey!" Goethe, *Jahrmarktsfest*, 265; trans. Denton, "The Technological Eye," 261.

36. *The Morning Chronicle* (January 5, 1797).

37. Personal communication with Francesco Modolo of the Museo del Precinema, Collezione Minici Zotti, which includes an eighteenth-century magic lantern with original slides.

38. Bauman, *North German Opera*, 206. The text performed in 1778 was an expanded version of the one Goethe originally published in 1773. The Schattenspielmann's text, however, remained unchanged from the 1773 version.

39. The music is preserved in Weimar, Herzogin Anna Amalia Bibliothek, Mus II a:2.

40. The music is repeated for each refrain, alternating the key of D (shown in example 1) and the same music transposed to the key of G.

41. "Natural history or geogony" singles out those numbers describing the formation of the earth, and its plant and animal life, namely Uriel's accompanied recitative no. 3, "And God made

the firmament"; Raphael's aria no. 6, in which mountains and rocks emerge from the waters, rivers snake through the plains, and brooks flow through valleys; Gabriel's aria no. 8, which introduces plants; Gabriel's aria no. 15, which adds birds; the trio no. 18, which mingles topology with plants, birds, and fish; and of course Raphael's accompanied recitative no. 21, which presents a series of mammals and insects.

42. Temperley, *Haydn*, 67.

43. Gabriela von Baumberg, quoted in Lawrence Kramer, *Classical Music and Postmodern Knowledge* (Berkeley: University of California Press, 1995), 74 (translation modified).

44. Nicholas Mathew, "Beethoven's Political Music, the Handelian Sublime, and the Aesthetics of Prostration," *19th-Century Music* 33 (2009): 116. Richard Will discusses the critical strategy of casting composers of tone-paintings as creators in *Characteristic Symphony*, 153–154.

45. Johann Gottfried Herder, *Shakespeare*, trans. Gregory Moore (Princeton, NJ: Princeton University Press, 2008), 53. The essay was first published in 1773.

46. "Wer sollte es wohl glauben! . . . Sie haben ihre Maschinen in Bewegung zu setzen gelernt; weiter frage man sie nicht." Fr. Von Beaumont, "Die antwort der Fr. Von Beaumont," *Allgemeines Magazin der Natur, Kunst und Wissenschaften*, 12 vols. (Leipzig: Gleditsch, 1756), 8:116.

47. [Jean-Pierre Claris de] Florian, *Fables de M. de Florian* (Paris: P. Didot, 1792), 79. On this story and its iconographic tradition, see Helen Weston, "In the Eye of the Artist," in *Realms of Light*, ed. Richard Crangle, Mervyn Heard, and Ine Van Dooren (London: Magic Lantern Society, 2005), 73–74; and Rossell, *Laterna magica*, 119–120.

48. For further discussion of Amalia's other numbers, see Deirdre Loughridge, "Haydn's *Creation* as an Optical Entertainment," *Journal of Musicology* 27 (2010): 43.

49. Triest, "Remarks on the Development of the Art of Music in Germany," 374

50. Ibid., 368–369.

51. Benjamin Martin, *The Young Gentleman and Lady's Philosophy*, 2nd ed., vol. 2 (London: W. Owen, 1772), 285.

52. Ibid., 286.

53. ". . . einen besonders schönen mit vielen Fleiß und Kösten verfertigten Schattenspiel, Prospekten, und mechanischen Feuerwerke die Ehre haben aufzuwarten . . . Da der Künstler weder Mühe noch Kosten gesparrt, um seine Vorstellungen nach dem Geschmack des Hochschätzbaren Publikums einzurichten, so schmeichlet er sich desto mehr eines zahlreichen Zuspruches, wo er versichert ist, daß gewiß Niemand ohne einer vollkommenen Zufriedenheit seinen Schauplaz verlassen wird." Broadside held by the Wienbibliothek im Rathaus, Plakatsammlung, Teil 1. D 64.522, I.N. 96.022.

54. See Deirdre Loughridge, "Celestial Mechanisms: Adam Walker's Eidouranion, Celestina, and the Advancement of Knowledge," in *Sound Knowledge: Music and Science in London, 1789–1851*, ed. J. Q. Davies and Ellen Lockhart (Chicago: University of Chicago Press, forthcoming). Thomas Tolley discusses Haydn's *Creation* in connection with Loutherbourg's Eidophusikon (which was no longer on view when Haydn visited London) in *Painting the Cannon's Roar*, 292–304.

55. Johann Heinrich Helmuth, *Volksnaturlehre zur Dämpfung des Aberglaubens*, 4th ed. (Braunschweig: Verlag der Schulbuchhandlung, 1798), 68; trans. in Hermann Hecht, *Pre-Cinema History: An Encyclopaedia and Annotated Bibliography of the Moving Image Before 1896* (London: Bowker-Saur, 1993), 53.

56. Barbara Maria Stafford, *Artful Science* (Cambridge, MA: MIT Press, 1996), xxi.

57. See Simon Schaffer, "Natural Philosophy and Public Spectacle in the Eighteenth Century," *History of Science* 21 (1983): 1–43.

58. Giuseppe Pinetti, *Physical Amusements and Diverting Experiments* (London, 1784), title page; "Theatralische Gaukelspiele in Berlin," *Journal des Luxus und Moden*, 422.

59. Simon Werrett, *Fireworks: Pyrotechnic Arts and Sciences in European History* (Chicago: University of Chicago Press, 2010).

60. Broadside held by the Wienbibliothek im Rathaus, Plakatsammlung, Teil 1. D 64.522, I.N. 96.022.

61. "Dann gab es mehrere öffentliche Gärten, wo fast in allen Tagen der Woche von den

Regiments- und Stadtmusikern Konzerte bei Bier und Tabak aufgeführt wurden, die ich nach und nach alle kennenlernte und fleißig besuchte, auch daselbst Klavier- und Violinkonzerte spielte." Carl Friedrich Zelter, *Darstellungen seines Lebens*, ed. Johann-Wolfgang Schottländer (New York: Georg Olms Verlag, 1978), 40.

62. Wilhelm Rintel identifies "Stadtumusikus George" as Lorenz George in *Carl Friedrich Zelter: Eine Lebensbeschreibung*, ed. Wilhelm Rintel (Berlin: Otto Janke, 1861), 35. Applegate, however, identifies him as Johann Friedrich George in *Bach in Berlin*, 129.

63. "In vier bis fünf großen Stuben waren die Wände mit den üblichen musikalischen Instrumenten bekleidet. Mitten in der Wohnstube stand ein Familientisch. An den Seiten wenige Stühle, eine Drechselbank mit Zubehör. Rapiere, Flinten, Reitzeug, Axt, Säge; Nutzhölzer traten unter dem Ofen und Bette hervor; Feuerwerksgeräte, auch eine Elektrisiermaschine fehlte nicht, und hundert Dinge, die man selten beisammen sieht." Zelter, *Darstellungen seines Lebens*, 41.

64. "Da George sich auch mit dem Feuerwerkswesen beschäftigte, wobei jeder seiner Freunde helfen, Papier, Pulver und dergleichen anschaffen mußte, so gab es auch dann und wann ein Feuerwerk, bei dessen Abbrennung geblasen und die Pauken gerührt wurden. Er nannte dieses sein Augenkonzert und konnte sich lange vorher kindlich auf den Augenblick freuen, alle diese Dinge anzuzünden und in die Luft spielen zu sehn. Es wurden papierne Drachen von ungemeiner Größe verfertigt, mit Kunstfeuern illuminiert und im Herbste auf dem Stoppelfelde gegen Abend mit langem Feuerschweife in die Luft gezogen. Der Jubel dabei, wenn alles wohl geriet, war erfreulich und viele Tage nachher der Gegenstand der Unterhaltungen." Ibid., 42.

65. "Den Ton der Sänger sah ich gleichsam kommen, doch das Orchester im ganzen war mir ein ungeheures, angenehmes Rätsel. Ich war mitten unter den Musikern, von denen jeder *ein* Instrument spielte, und doch hörte ich nicht *eins*, sondern das Orchester selber, welches ich mir wie einen bezauberten Kasten, wie eine Art von Orgel vorstellte, tönte und klang als *ein* Ganzes in mir wider. Von dieser Betrachtung an ward mein Blick auf das Theater gelockt, und ich schwamm in einem Meer von Freuden." Ibid., 16.

66. Carl Friedrich Zelter, "Briefe an einen Freund über die Musik in Berlin," *Allgemeine musikalische Zeitung* 3 (January 21, 1801), col. 293; trans. in Landon, *Haydn: Chronicle and Works*, 4:588.

67. Carl Friedrich Zelter, "Die Jahreszeiten nach Thomson," *Allgemeine musikalische Zeitung* 6 (May 2, 1804), col. 515; trans. in Landon, *Haydn: Chronicle and Works*, 5:189.

68. Zelter, "Briefe an einen Freund über die Musik in Berlin," col. 294; trans. in Landon, *Haydn: Chronicle and Works*, 4:588.

69. Zelter, "Die Jahreszeiten nach Thomson," col. 515–516; trans. in Landon, *Haydn: Chronicle and Works*, 5:189.

70. Donald Tovey, *Essays in Musical Analysis* (London: Oxford University Press, 1981), 350; Dolan, *Orchestral Revolution*, 136–148.

71. Landon, *Haydn: Chronicle and Works*, 4:587.

72. Zelter, "Die Schöpfung," col. 391; trans. in Temperley, *Haydn*, 90.

73. Zelter, "Briefe an einen Freund über die Musik in Berlin," 293; trans. in Landon, *Haydn: Chronicle and Works*, 4:588.

74. Landon, *Haydn: Chronicle and Works*, 4:455.

75. MacIntyre, *Haydn*, 283.

76. Georg August Griesinger, *Biographische Notizen über Joseph Haydn* (Leipzig, 1810), trans. by Vernon Gotwals as *Joseph Haydn: Eighteenth-Century Gentleman and Genius* (Madison: University of Wisconsin Press, 1963), 49. According to MacIntyre, the gesture was first reported by L. von Seckendorft, "Kunstnachrichten aus Wien: Musik," *Prometheus* 1, no. 3 (1808): 17; see MacIntyre, *Haydn*, 263, n313.

77. Helmuth, *Volksnaturlehre zu Dämpfung des Aberglaubens*, 53.

78. See Matthew Head, "Music with 'No Past'? Archaeologies of Joseph Haydn and 'The Creation'," *19th-Century Music* 23 (2000): 196, 206.

79. Triest, "Remarks on the Development of the Art of Music in Germany," 368–369. Triest's experience of *The Creation* may thus be seen to anticipate Georges Duhamel's reaction to cinema, cited by Walter Benjamin in his mechanical-reproduction essay to illustrate the "shock effect" of

its changing scenes and their incompatibility with contemplation: "I can no longer think what I want to think. My thoughts have been replaced by moving images." Walter Benjamin, *Illuminations* (New York: Schocken Books, 1978), 238.

80. Quoted in Stephanie Campbell, "Seeing Music: Visuality in the Friendship of Johann Wolfgang von Goethe and Carl Friedrich Zelter," in *The Arts Entwined*, ed. Marsha Morton and Peter L. Schmunk (New York: Garland Publishing, 2000), 50; see Campbell's essay for further context and analysis of Zelter's audiovisuality.

81. "Eine neue Art von Concerten wird diesen Winter in Paris die Modewelt unterhalten. Alles, was die Musik ausdrückt, soll während derselben durch Transparentgemäldge in trefflicher Ausführung zur Anschauung gebracht werden. Haydn's Schöpfung wird einfludirt und muß, von passenden Phantasmagorien begleitet, die Sinne der Zuhörer doppelt umstricken." August Lewald, *Album der Boudoirs* (Leipzig, 1836), 42–43; trans. in Walter Benjamin, *The Arcades Project*, ed. Rolf Tiedemann (Cambridge, MA: Harvard University Press, 1999), 530.

82. "Récréations Musicales et Pittoresques," *Le Ménestrel* 158 (December 11, 1836): [1–2]. Porro's Nocturnorama was considered a failure, its run ending after only fifteen days. However, the apparatus was revived the following year by conductor Louis Antoine Jullien, who presented seven transparent pictures with his own newly composed work, *Balthasar's Feast. Le Ménestrel* (August 13, 1837): [1].

83. For Berlioz (writing in 1857), Haydn's creation of light called to mind the Carcel lamp, which though patented in 1800 remained among the brightest generally available. Quoted in Temperley, *Haydn*, 43.

84. "Mehr jedoch als zu diesem großartigen Werke scheint mir diese Einrichtung zu heitern und senimtentalen Anregungen geeignet." August Lewald, *Album der Boudoirs*, 43; trans. in Benjamin, *The Arcades Project*, 530.

85. Lewald reported that Haydn's *Creation* had a "magical effect" in Toulouse, in contrast to Paris, where romances and contradanses dominated and Haydn's symphonies no longer made any effect. August Lewald, "Musik," in *Europa, chronic der gebildeten welt*, vol. 1 (Leipzig and Stuttgart: J. Scheible's Verlags-Expedition, 1835), 283.

86. Lawrence Kramer, "Music and Representation: The Instance of Haydn's *Creation*," in *Music and Text: Critical Inquiries*, ed. Steven Paul Scher (Cambridge: Cambridge University Press, 1992), 139.

CHAPTER FIVE

1. Terry Castle, "Phantasmagoria: Spectral Technology and the Metaphorics of Modern Reverie," *Critical Inquiry* 15 (1988): 26–61.

2. Theodor Adorno, *In Search of Wagner* (London: NLB, 1981), 86.

3. *The Critical Reception of Beethoven's Compositions by His German Contemporaries*, ed. Wayne M. Senner, Robin Wallace, and William Meredith, 2 vols. (Lincoln: University of Nebraska Press, 1999), 2:8.

4. Crary, *Techniques of the Observer*, 23.

5. Immanuel Kant, "Dreams of a Spirit-Seer Elucidated by Dreams of Metaphysics (1766)," in *Theoretical Philosophy, 1755–1770*, ed. David Walford (Cambridge: Cambridge University Press, 1992), 339.

6. Ibid., 327.

7. Ibid., 354.

8. Johann Gottfried Herder, "Emanuel Swedenborg: The Greatest Spirit-Seer of the Eighteenth Century," trans. Gregory R. Johnson and Glenn Alexander Magee, *Studia Swedenborgiana* 14 (2005) [online].

9. "Die erste ist die künstliche, die in optischem Betruge besteht. Die zweyte is die, die durch Bilder der Einbildungskraft erzeugt wird, da die Einbildung ein Bild ausser sich schafft. Und die dritte ist die wahre Geistererscheinung, die nur dem innern Sinne sichtbar ist, und durch eben diesen innern Sinne den äussern Sinnen zum Bild geschaffen wird, welches letzte die wahre Erscheinung ist." Karl von Eckartshausen, *Aufschlüsse zur Magie aus geprüften Erfahrungen über verbor-*

gene philosophische Wissenschaften und verdeckte Geheimnisse der Natur, vol. 2 (Munich: Joseph Leutner, 1791), 65. Trans. Andriopoulos, *Ghostly Apparitions*, 36.

10. N. W. Wraxall, *Memoirs of the Courts of Berlin, Dresden, Warsaw, and Vienna in the Years 1777, 1778 and 1779*, vol. 1 (London: Cadell, 1800), 278.

11. Johann Samuel Traugott Gehler, *Physikalisches Wörterbuch oder Versuch einer Erklärung der vornehmsten Begriffe und Kunstwörter der Naturlehre*, vol. 3 (Leipzig: Schwickert, 1790), 89; see chapter 4.

12. "Diese Antwort heilte mich von meinem Schrecken, denn ich konnte ganz deutlich unterscheiden, daß der Geist nicht hebräisch, sondern jüdisch, das ist, in korrupten hebräischen Wörtern mit untermischten deutschen Bindewörtern antwortete; die himmlischen Geister werden doch wohl rein hebräisch reden." Christlieb Benedict Funk, *Natürliche Magie oder Erklärung verschiedner Wahrsager- und Natürlicher Zauberkünste* (Berlin und Stettin: Friedrich Nicolai, 1783), 262.

13. ". . . man höre seine heisere unnatürliche Stimme: sogleich verschwindet unsre Philosophie, und wir zittern im Kreise, anstatt daß wir über die gerunzelte Stirn des Magus lachen sollten." Johann Samuel Halle, *Magie oder die Zauberkräfte der Natur, so auf den Nutzen und die Belustigung angewandt worden*, vol. 1 (Wien: Johann Thomas Edlen von Trattnern, 1787), 244.

14. "Es entsteht vorher ein Gepolter, und nun ist der Zuschauer mehr todt als lebendig, seine Einbildungskraft ist nun äußerst erregt, und geschickt jede Eindrücke anzunehmen, welcher der Zauberer ihr geben will." Johann Peter Eberhard, *Abhandlungen vom physikalischen Aberglauben und der Magie* (Halle: Magdeburgischen, 1778), 100.

15. "Die angeblichen Magie machen vorher solche Zurüstungen, wodurch die Zuschauer in Furcht und Schrecken gesetzt, und daher unfahig werden, den Betrug zu merken." Ibid., 99.

16. Christoph Friedrich Nicolai, "A Memoir on the Appearance of Spectres or Phantoms occasioned by Disease, with Psychological Remarks. Read by Nicolai to the Royal Society of Berlin, on the 28th of February, 1799," *A Journal of Natural Philosophy, Chemistry, and the Arts* 6 (November 1803): 176. Nicolai's report included an infamous cure for his visions—application of leeches to the anus—which Goethe, Hoffmann, and others mocked in their writings.

17. von der Reck, "Nachricht von der Philidorschen Geisterbeschwörung," *Berliner Monatsschrift* 13/5 (May 1789): 456–473; translated in Mervyn Heard, *Phantasmagoria: The Secret Life of the Magic Lantern* (Hastings: Projection Box, 2006), 65–69.

18. "Wir wünschen übrigens jedem orte in Deutschland, wo dergleichen Gaukelein getrieben werden, eine Gesellschaft von solchen umbefangenen unleichtgläubigen Zuschauers und von solchen einsichtsvollen wahrheitliebenden Forschern." J. v. Schwarzkopf, "Erzählung einer neulichen Geisterzitation in Berlin," *Berliner Monatsschrift* 13/5 (May 1789): 474.

19. von der Reck, "Nachricht von der Philidorschen Geisterbeschwörung," 473; translated in Heard, *Phantasmagoria*, 69.

20. "Selbst eine Pfuscherei in Swedenborgs, Schröpfers und Kagliostros Handwerk liesert einen nützlich Beitrag zur Geschichte der modernen geheimen Künste: theils in sofern sie die Mannichfaltigkeit oder den Mechanismus des Betruges aufdekt; theils, in sofern sich aus dem Betragen der Zuschauer die Stimmung und Empfänglichkeit der Ortsbewohner für solche Alfanzereien beurtheilen läßt." Schwarzkopf, "Erzählung einer neulichen Geisterzitation in Berlin," 474.

21. These titles come from three broadsides, dating from 1790–1791, held by the Wienbibliothek im Rathaus, Plakatsammlung, Teil 1. D 64.522, I.N. 96.022.

22. Cited in Mannoni, *Great Art of Light and Shadow*, 150.

23. "Herr Phylidor seit langer Zeit beschäftiget der künstlichen Geister-Erscheinung mehrere Vollkommenheit zu geben, hat nach vielen Versuchen eine ganz neue Methode erfunden . . ." Broadside titled "Phantasmorasi" in Wienbibliothek im Rathaus, Plakatsammlung, Teil 1. D 64.522, I.N. 96.022.

24. ". . . höret man von ferne den Donner sich näheren, der sich nach und nach immer mehr verstärkt, und mit Hagel, und brausendem Wind begleitet wird . . ." Broadside titled "Schröperischen, und Cagliostroischen Geister-Erschinungen," in ibid.

25. ". . . es erlöscht ein Licht nach dem andern von sich selbst, und es steigen aus den Lampen

feurige Flammen empor, hierauf erlöschen die Lampen, und man ist sodann in einer undurchdringlichen Finsterniß . . ." Ibid.

26. ". . . verschiedene Geister erblickt nun das Auge des Zuschauers von aller Gestalt und Größe, sie flattern in einem Zirkel herum, und verschwinden wieder . . ." Ibid.

27. ". . . in jeder Vorstellung erscheint die Auhnlichkeit verschiedener der Gesellschaft sehr bekannter Personen in der Gestalt des so allgemein angenommenen Begriffes eines Geistes, ein jeder von Ihnen entsteht, und verschwindet auf eine andere Art, einer kommt aus dem Fußboden empor; der Zweyte stehet plötzlich auf einmal da, der dritte bildet sich aus einer grauen Wolke, und wird nach und nach so körperlich, daß man deutlich die vorgebliche Person in der Gestalt erkennt. Eine entsteht in der Entfernung, und so täuschend, als wenn selbe 200 Schritt weit entfernt wäre, und nähere sich bis zum greifen, jede Erscheinung macht einige Schritt vorwärts gegen die Zuschauer und verschwindet." Ibid.

28. Recalling the "phantasmagorical manner" in which the Spirit appeared in the first scene of *Faust* in an 1819 performance, Goethe elaborated: "that is, in a darkened theater an illuminated head is projected from the rear upon a screen stretched across the background, first as a small image, then gradually increasing in size, so that it seems to be coming closer and closer." Trans. in Frederick Burwick, *Illusion and the Drama: Critical Theory of the Enlightenment and Romantic Era* (University Park: Pennsylvania State University Press, 1991), 241.

29. On stories of early cinematic spectators fleeing film of an approaching train—the significance of which has been contested—see James Lastra, *Sound Technology and the American Cinema: Perception, Representation, Modernity* (New York: Columbia University Press, 2000), 64–65.

30. Mannoni, *Great Art of Light and Shadow*, 162.

31. In her landmark 1988 article on the then largely forgotten origins of the term *phantasmagoria* in late eighteenth-century ghost shows, Terry Castle repeated Robertson's claim to have invented the phantasmagoria. Laurence Mannoni and Mervyn Heard have since recovered the prior art of Phylidor. Enslen, however, remains absent from most histories of phantasmagoria, likely as a result of the comparatively scant material on his career, his preference for the name "optical presentation," and his seeming irrelevance from the perspective of technological progress. Castle, "Phantasmagoria"; Mannoni, *Great Art of Light and Shadow*; Heard, *Phantasmagoria*.

32. ". . . weder Fleiß, Mühe, noch Kosten gesparret ein mechanisch musikalisches Theater zu verfertigen, worin die Musik, die Maler- und Bildhauerkunst mit der Mechanik verbunden, die Hauptgegenstände ausmachen, die in Gestalt lebendscheinender Automate, oder künstliche beweglicher Figuren erscheinen . . . Zuschauer überraschen und auf das angenehmste unterhalten werden." *Wiener Zeitung* (December 7, 1791): 3131.

33. The latest announcement of Enslen presenting his "mechanical and optical art and moving spirit-appearances, together with an optical ballet" I have found comes from the *Swäbischer Mercur* for May 19, 1803, when Enslen ended a performance run in Stuttgart. According to Franz Josef Pisko, Enslen continued to offer phantasmagoria to the Berlin public until about 1833. According to Deac Rossell, however, Enslen retired to an estate near Danzig from 1805 to 1807, in which year the estate was confiscated by Napoleon's troops and Enslen resumed life as a traveling showman, now exhibiting panoramas with his son. [Mechanische und optische Kunstwerke], *Swäbischer Mercur* (99), May 19, 1803, 465. Franz Josef Pisko, *Licht und Farbe: Eine gemeinfassliche Darstellung der Optik*, vol. 2 (Munich: R. A. Oldenbourg, 1869), 159. Deac Rossell, "Enslen, Johann Carl (1759–1848)," in *Encyclopedia of Nineteenth-Century Photography*, vol. 1, ed. John Hannavy (New York: CRC Press, 2007), 492.

34. "Nun wurden alle Lichter im ganzen Schauspielhause ausgelöscht, dichte Vorhänge vor den Fenstern sielen herab, die Logen-Thüren wurden, dem Tageslicht allen Zugang zu verspeiren, verschloßen, und dumpfe Donner rollten fürchterlich daher. Das Vorgefühl der Geistererscheinungen ward rege gemacht, Schaudern ergriff den Furchtsamen, und Geister ließen sich sehen." J. W. A. Kosmann, "Ueber das Kunstkabinet und die optischen Vorstellungen des Herrn Enslen," in *Denkwürdigkeiten und Tagesgeschichte der Mark Brandenburg*, vol. 2, ed. J. W. A. Kosmann and Th. Heinsius (Berlin: Beliz und Braun, 1797), 730. The *Journal des Luxus und Moden* remarked only:

"some thunder from behind the scenes abrogated the appearance of a floating spirit-figure" (Ein Donner hinter den Koulissen kündigte zuerst die Erscheinung einer schwebenden Geistesgestalt an). "Theatralische Gaukelspiele in Berlin: Berlin den 20ten Juni 1796," *Journal des Luxus und der Moden*, 11 (1796): 427.

35. "Theatralische Gaukelspiele in Berlin: Berlin den 20ten Juni 1796," 428.

36. Kosmann, "Ueber das Kunstkabinet und die optischen Vorstellungen des Herrn Enslen," 744.

37. "Doch übertrift Hr. Enslens Kunst das Gewöhnliche sehr weit, und seine Geisterfabrik würd in England ohne Bedenken ein Patentartikel. Den 24th Junius beehrte der König selbst mit den Prinzen und Prinzessin sein Theater. Es wurde diesmal anstatt der Erscheinung Friedrich II, ein transparentes Bild des Königs vorgestellt. Das Orchester spielte das Lied: Heil dir im Siegeskranze! und das ganze Parterre sang es laut." "Theatralische Gaukelspiele in Berlin: Berlin den 20ten Juni 1796," 429. According to another report, "the folksong 'Heil, Heil Herrscher Dir'" was sung by a choir and the audience joined in. Kosmann, "Ueber das Kunstkabinet und die optischen Vorstellungen des Herrn Enslen," 744.

38. Mozart, *The Letters of Mozart and His Family*, vol. 2, 683.

39. Quoted in Mannoni, *Great Art of Light and Shadow*, 144.

40. "Die Kenntniß von dem versteckten natürlichen Zusammenhange wird seine Neugierde befriedigen und ihn dabey vergnügen. Dieß wird noch dabey den Nutzen haben, daß man Gaukeleyen und Taschenspielereyen nicht vor unbekannte Kräfte der Natur ausgeben wird. Und wenn sollte dieß nöthiger seyn, als zu unserer Zeit, die einen Schröpfer, Gaßner, Cagliostro, die Wunderkräfte des Magnetisums, Güldenfalts alchemische Mährchen, und die Offenbarungen des Schwedenborgischen neuen Jerusalems zeugte." Wiegleb, *Die natürliche Magie*, 4.

41. E. T. A. Hoffmann, *The Best Tales of Hoffmann*, ed. E. F. Bleiler (New York: Dover Publications, 1967), 81.

42. E. T. A. Hoffmann, *Fantasy Pieces in Callot's Manner: Pages from the Diary of a Traveling Romantic*, trans. Joseph M. Hayse (Schenectady, NY: Union College Press, 1996), 233.

43. Letter to Julius Eduard Hitzig (April 28, 1812), trans. in Hoffmann, *E. T. A. Hoffmann's Musical Writings*, 64.

44. "Theater und Musik: Ueber die Aufführung Calderonischer Stücke auf dem Theater in Bamberg (aus Briefen)," *Journal des Luxus und Moden* 27 (January 1812): 28, http://zs.thulb.uni-jena.de/receive/jportal_jparticle_00094415.

45. "Seine Symphonie führt uns in unabsehbare, grüne Hayne, in ein lustiges, buntes Gewühl glücklicher Menschen. Jünglinge und Mädchen schweben in Reihentänzen vorüber . . . Ein Leben voll Liebe, voll Seligkeit, wie vor der Sünde, in ewiger Jugend; kein Leiden, kein Schmerz; nur süsses, wehmüthiges Verlangen nach der geliebten Gestalt, die ferne im Glanz des Abendrothes daher schwebt, nich näher kömmt und nicht verschwindet." [E. T. A. Hoffmann], "Recension," *Allgemeine musikalische Zeitung* 12, July 4, 1810, col. 632; trans. Bonds, *Music as Thought*, 34, and Hoffmann, *E. T. A. Hoffmann's Musical Writings*, 237.

46. "Glühende Strahlen schießen durch dieses Reiches tiefe Nacht, und wir werden Riesenschatten gewahr, die auf- und abwogen, enger und enger uns einschließen, und alles in uns vernichten, nur nicht den Schmerz der unendlichen Sehnsucht, in welcher jede Lust, die, schnell in jauchzenden Tönen emporgestiegen, hinsinkt und untergeht, und nur in diesem Schmerz, der, Liebe, Hoffnung, Freude in sich verzehrend, aber nicht zerstörend, unsre Brust mit einem vollstimmigen Zusammenklange aller Leidenschaften zersprengen will, leben wir fort und sind entzückte Geisterseher." Hoffmann, "Recension," col. 633.

47. "Review of Beethoven's Fifth Symphony," in Hoffmann, *E. T. A. Hoffmann's Musical Writings*, 240.

48. *Allgemeine musikalische Zeitung*, 14, August 5, 1812, col. 519–526; trans. as "Review of Beethoven's Overture to *Coriolan*," in Hoffmann, *E. T. A. Hoffmann's Musical Writings*, 287.

49. Senner, Wallace, and Meredith, *The Critical Reception of Beethoven's Compositions*, 2:103.

50. Hoffmann, *E. T. A. Hoffmann's Musical Writings*, 303.

51. Scott Burnham, *Beethoven Hero* (Princeton, NJ: Princeton University Press, 1995), 32.

52. Hoffmann, *E. T. A. Hoffmann's Musical Writings*, 247.

53. Ibid., 250.

54. Heinrich Schenker, *Der Tonwille: Pamphlets/Quarterly Publications in Witness of the Immutable Laws of Music*, vol. 2, ed. William Drabkin (New York: Oxford University Press, 2004 [1921–1924]), 26.

55. Dahlhaus, *The Idea of Absolute Music*, 69.

56. Dahlhaus, *Nineteenth-Century Music*, 95.

57. Ibid., 91.

58. Burnham, *Beethoven Hero*, 150–151.

59. Lawrence Kramer, *Music and Poetry, the Nineteenth Century and After* (Berkeley: University of California Press, 1984), 240–241; Kinderman, *Beethoven*, 153. Kinderman borrows the phrase "C major that cannot be followed" from the analysis by Kramer just cited.

60. Hoffmann, *E. T. A. Hoffmann's Musical Writings*, 250.

61. The supernatural, ghoulish character of the third movement has been widely recognized; Donald Tovey described the pizzicato statement of the main theme as "one of the ghostliest things ever written, with something of the thin, bickering quality of the poor ghosts that Homer describes where Odysseus visits the Land of Shadows." Cited in Michael Broyles, *Beethoven: The Emergence and Evolution of Beethoven's Heroic Style* (New York: Excelsior Music, 1987), 195.

62. Edward Morgan Forster, *Howards End* (New York: A. A. Knopf, 1921), 40–41.

63. Burnham identifies "the feeling that when one listens to Beethoven's music one is in the presence of something more than music" with Beethoven's heroic style, this presence becoming "uncanny" in the Fifth. Burnham, *Beethoven Hero*, 147, 163–164.

64. Dahlhaus, *Nineteenth-Century Music*, 91.

65. Chantler, *E. T. A. Hoffmann's Musical Aesthetics*, 67.

66. Voskuhl, *Androids in the Enlightenment*, 179.

67. Newcomb, "New Light(s) on Weber's Wolf's Glen Scene"; Thomas Grey, "Tableaux Vivants: Landscape, History Painting, and the Visual Imagination in Mendelssohn's Orchestral Music," *19th-Century Music* 21 (1997): 38–76.

CONCLUSION

1. Leon Botstein, "Music, Technology, and the Public," *Musical Quarterly* 78/2 (1994): 178, 181.

2. Arved Mark Ashby, *Absolute Music, Mechanical Reproduction* (Berkeley: University of California Press, 2010), 246–248.

3. "Music Discovery Still Dominated by Radio, Says Nielsen Music 360 Report," http://www.nielsen.com/us/en/press-room/2012/music-discovery-still-dominated-by-radio—says-nielsen-music-360.html (August 14, 2012); Ben Sisario, "What's Billboard's No. 1? Now YouTube Has a Say," *New York Times*, February 21, 2013, C3.

4. Alex Ross, "Schubert on the Beach: The New World Symphony's Radical New Home," *The New Yorker* (February 14, 2011) [online].

5. Carol Vernallis, *Unruly Media: YouTube, Music Video, and the New Digital Cinema* (Oxford: Oxford University Press, 2013); *Oxford Handbook of New Audiovisual Aesthetics*, ed. John Richardson, Claudia Gorbman, and Carol Vernallis (Oxford: Oxford University Press, 2013); *Oxford Handbook of Sound and Image in Digital Media*, ed. Carol Vernallis, Amy Herzog, and John Richardson (Oxford: Oxford University Press, 2013). Though none of these volumes elaborates on the meaning of "audiovisual turn," the phrase echoes W. J. T. Mitchell's "pictorial turn" and shares with it a problematic implication. As Mitchell notes, his phrase has often been taken to mean that "the modern era is unique or unprecedented in its obsession with vision and visual representation." Mitchell instead meant the "pictorial turn" to identify a trope, recurrent since antiquity, that credits (or blames) a new state of vision or images for bringing about a historical turning point. Hence, "a critical and historical use of this figure would be as a diagnostic tool to analyze specific moments when a new medium, a technical invention, or a cultural practice erupts in symptoms of panic or euphoria (usually both) about the visual." Mitchell, "Showing Seeing," 173.

6. The volume *Multimedia Histories* dates "the proliferation of visual and audio technologies"

to the early nineteenth century; *Multimedia Histories: From the Magic Lantern to the Internet*, ed. James Lyons and John Plunk (Exeter: University of Exeter Press, 2007), xvii. Wagner's concept of the Gesamtkunstwerk represents another favorite starting point for histories of multimedia; see *Multimedia: From Wagner to Virtual Reality*, ed. Randall Packer and Ken Jordan (New York: W. W. Norton & Co., 2002).

7. Claudia Gorbman, *Unheard Melodies: Narrative Film Music* (Bloomington: Indiana University Press, 1987); Brian Kane, *Sound Unseen*.

8. "Doch mochte ich ihn nur in dem seinigen, wo der Brennpunkt des Ganzen er selbst war, sehen und hören. Ich sage mit Bedacht: *sehen* und hören. Die allgemeine Begierde, im Concert nicht allein zu hören, sondern auch zu sehen, das Drängen nach Plätzen im Saal, wo dies möglich ist, entsteht gewiss nicht aus blosser, müssiger Schaulust: man hört besser, wenn man sieht; die geheime Verwändschaft von Licht und Ton offenbart sich deutlich; beydes, Licht und Ton, gestaltet in individueller Form, und so wird der Solospieler, die Sängerin, selbst die ertönende Melodie!" *Allgemeine musikalische Zeitung*, 17 (January 11, 1815): col. 19; "Letters on Music in Berlin," 389–390. The October 23, 1814, program comprised all music by Romberg: a symphony, his Sixth Cello Concerto (militaire), and his Rondoletto and Capriccio on Swedish folk songs.

9. Le Guin, *Boccherini's Body*; Voskuhl, *Androids in the Enlightenment*.

10. Ross, "Schubert on the Beach."

11. Ibid., 20.

12. James Buhler and David Neumeyer suggested how different music history would look if the modernist commitment to evolutionary progress and suspicion of mass culture were abandoned and film music were allowed to offer an account of late Romantic style. James Buhler and David Neumeyer, "*Strains of Utopia: Gender, Nostalgia, and Hollywood Film Music* by Caryl Flinn; *Settling the Score: Music and the Classical Hollywood Film* by Kathryn Kalinak" [review], *Journal of the American Musicological Society* 47 (1994): 364–385.

SELECT BIBLIOGRAPHY

Abbate, Carolyn. *In Search of Opera*. Princeton, NJ: Princeton University Press, 2001.

Abel, Richard, ed. *Encyclopedia of Early Cinema*. New York: Routledge, 2005.

Abraham, Gerald, ed. *The Age of Beethoven, 1790–1830*. New York: Oxford University Press, 1982.

Adelung, Johann Christoph. *Ueber den deutschen Styl*. Berlin: C. F. Voss, 1785.

Adorno, Theodor. *In Search of Wagner*. London: NLB, 1981.

———. "Bourgeois Opera." In *Sound Figures*, trans. Rodney Livingstone, 15–28. Stanford, CA: Stanford University Press, 1999.

Albright, Daniel. *Untwisting the Serpent: Modernism in Music, Literature, and the Other Arts*. Chicago: University of Chicago Press, 2000.

Allanbrook, Wye Jamison, ed. *Strunk's Source Readings in Music History: The Late Eighteenth Century*. New York: W. W. Norton, 1998.

Altick, Richard Daniel. *The Shows of London*. Cambridge, MA: Belknap Press, 1978.

Altman, Rick. "Moving Lips: Cinema as Ventriloquism." *Yale French Studies* 60 (1980): 67–79.

———. *Silent Film Sound*. New York: Columbia University Press, 2007.

Andriopoulos, Stefan. *Ghostly Apparitions: German Idealism, the Gothic Novel, and Optical Media*. New York: Zone Books, 2013.

Applegate, Celia. *Bach in Berlin: Nation and Culture in Mendelssohn's Revival of the St. Matthew Passion*. Ithaca, NY: Cornell University Press, 2005.

Armstrong, Tim. *Modernism, Technology, and the Body: A Cultural Study*. New York: Cambridge University Press, 1998.

Ashby, Arved. *Absolute Music, Mechanical Reproduction*. Berkeley: University of California Press, 2010.

Astley, Philip. *The favourite airs for the violin, German flute, harp, or harpsichord, with a bass; together with the scene of the broken-bridge in the* Ombres chinoises. London, 1780.

Baker, Nancy Kovaleff, and Thomas Christensen, eds. *Aesthetics and the Art of Musical Composition in the German Enlightenment: Selected Writings of Johann Georg Sulzer and Heinrich Christoph Koch*. Cambridge: Cambridge University Press, 1995.

Bauman, Thomas. *North German Opera in the Age of Goethe*. Cambridge: Cambridge University Press, 1985.

Baumgarten, Alexander Gottlieb. *Mediationes philosophicae de nonnullis ad poema pertinentibus*. Halle: J. Grunerti, 1734.

Beall, Karen. *Kaufrufe und Strassenhändler: Eine Bibliographie/Cries and Itinerant Trades: A Bibliography*. Hamburg: Hauswedell, 1975.

Beaumont, Fr. Von. "Die antwort der Fr. Von Beaumont." *Allgemeines Magazin der Natur, Kunst und Wissenschaften* 8 (Leipzig: Gleditsch, 1756): 114–172.

Benjamin, Walter. *Illuminations*. New York: Schocken Books, 1978.

———. "Doctrine of the Similar (1933)," trans. Knut Tarnowski. *New German Critique* 17 (1979): 65–69.

———. *The Arcades Project*, ed. Rolf Tiedemann. Cambridge, MA: Harvard University Press, 1999.

———. "On the Concept of History." In *Benjamin: Selected Writings*, vol. 4, ed. Howard Eiland and Michael W. Jennings. Cambridge, MA: Harvard University Press, 2003.

Bennett, James. "The Mechanics' Philosophy and the Mechanical Philosophy." *History of Science* 24 (1986): 1–28.

Berger, Karol. *Bach's Cycle, Mozart's Arrow*. Berkeley: University of California Press, 2007.

Bergman, Gösta M. *Lighting in the Theatre*. Stockholm: Almqvist & Wiksell International, 1977.

Berlin, Isaiah. *Three Critics of the Enlightenment: Vico, Hamann, Herder*, ed. Henry Hardy. Princeton, NJ: Princeton University Press, 2013.

Blumenberg, Hans. *The Legitimacy of the Modern Age*, trans. Robert M. Wallace. Cambridge, MA: MIT Press, 1985.

Bodley, Lorraine Byrne. *Goethe and Zelter: Musical Dialogues*. Burlington, VT: Ashgate, 2009.

Boehn, Max von. *Puppets and Automata*, trans. Josephine Nicoll. New York: Dover, 1972.

Bonanni, Filippo. *Gabinetto armonico pieno d'istromenti sonori indicati e spiegati*. Rome: Giorgio Placho, 1722.

Bonds, Mark Evans. "Idealism and the Aesthetics of Instrumental Music at the Turn of the Nineteenth Century." *Journal of the American Musicological Society* 50 (1997): 387–420.

———. *Music as Thought: Listening to the Symphony in the Age of Beethoven*. Princeton, NJ: Princeton University Press, 2006.

———. *Listen to This*. Upper Saddle River, NJ: Pearson Prentice Hall, 2009.

Borchmeyer, Dieter. *Drama and the World of Richard Wagner*. Princeton, NJ: Princeton University Press, 2003.

Bordat, Denis, and Francis Boucrot. *Les théâtres d'ombres: histoire et techniques*. Paris: L'Arche, 1956.

Botstein, Leon. "Music, Technology, and the Public." *Musical Quarterly* 78 (1994): 177–188.

Brago, Michael. "Haydn, Goldoni, and *Il mondo della luna*." *Eighteenth-Century Studies* 17 (1984): 308–332.

Breene, Samuel. "The Instrumental Body in the Age of Mozart: Science, Aesthetics and Performances of the Self." *Early Music* 42 (2014): 231–247.

Breitkopf und Härtel. "Nachricht für Theaterdirektionen." *Allgemeine musikalische Zeitung* 1/11 (April 1799).

Brittan, Francesca. "On Microscopic Hearing: Fairy Magic, Natural Science, and the *Scherzo Fantastique*." *Journal of the American Musicological Society* 64 (2011): 527–600.

Broyles, Michael. *Beethoven: The Emergence and Evolution of Beethoven's Heroic Style*. New York: Excelsior Music, 1987.

Buch, David. *Magic Flutes & Enchanted Forests: The Supernatural in Eighteenth-Century Musical Theater*. Chicago: University of Chicago Press, 2008.

Buhler, James, and David Neumeyer. "*Strains of Utopia: Gender, Nostalgia, and Hollywood Film Music* by Caryl Flinn; *Settling the Score: Music and the Classical Hollywood Film* by Kathryn Kalinak" [review]. *Journal of the American Musicological Society* 47 (1994): 364–385.

Bürger, Gottfried August. *Lenore, Ballade von Bürger: in Drei Englischen Übersetzungen*. Göttingen: Johann Christian Dieterich, 1797.

———. *Bürger's Sammtliche Werke*, ed. August Wilhelm Bohtz. Göttingen: Dieterichschen, 1835.

———. *Briefe von und an Gottfried August Bürger*, ed. Adolf Strodtmann. Berlin: Gebrüder Paetel, 1874.

———. *Bürgers Werke in Einem Band*, ed. Lore Kaim-Kloock and Siegfried Streller. Weimar: Volksverlag, 1962.

———. "Ankündigung (einer Neubearbeitung von 'Tausend und eine Nacht')." In *Sämtliche Werke*, ed. Günter und Hiltrud Häntzschel. Munich: Carl Hanser Verlag, 1987.

"Bürgers Leonore von André in Musik gesezt und in Chinesischem Schattenspiel aufgeführt von der Fräulein von B*** zu Regensburg 1781." *Litteratur- und Theater-Zeitung* 37 (September 15, 1781): 577–583.

Burnham, Scott. *Beethoven Hero*. Princeton, NJ: Princeton University Press, 1995.

Burwick, Frederick. *Illusion and the Drama: Critical Theory of the Enlightenment and Romantic Era*. University Park: Pennsylvania State University Press, 1991.

Campbell, Stephanie. "Seeing Music: Visuality in the Friendship of Johann Wolfgang von Goethe and Carl Friedrich Zelter." In *The Arts Entwined*, ed. Marsha Morton and Peter L. Schmunk, 47–62. New York: Garland Publishing, 2000.

Castelvecchi, Stefano. "From Nina to Nina: Psychodrama, Absorption, and Sentiment in the 1780s." *Cambridge Opera Journal* 8 (1996): 91–112.

Castle, Terry. "Phantasmagoria: Spectral Technology and the Metaphorics of Modern Reverie." *Critical Inquiry* 15 (1988): 26–61.

———. *The Female Thermometer: Eighteenth-Century Culture and the Invention of the Uncanny*. New York: Oxford University Press, 1995.

Chantler, Abigail. *E. T. A. Hoffmann's Musical Aesthetics*. Aldershot: Ashgate, 2006.

Charlton, David. *Grétry and the Growth of Opéra-Comique*. Cambridge: Cambridge University Press, 1986.

———. *French Opera, 1730–1830: Meaning and Media*. Brookfield: Ashgate, 2000.

Cheesman, Tom. *The Shocking Ballad Picture Show: German Popular Literature and Cultural History*. Oxford: Berg Publishers, 1994.

Chen, Fan Pen Li. *Chinese Shadow Theatre: History, Popular Religion, and Women Warriors*. Montreal: McGill-Queen's University Press, 2007.

Chion, Michel. *Audio-Vision: Sound on Screen*, ed. and trans. Claudia Gorbman. New York: Columbia University Press, 1994.

Christensen, Thomas. "Public Music in Private Spaces: Piano-Vocal Scores and the Domestication of Opera." In *Music and the Cultures of Print*, ed. Kate van Orden, 67–93. New York: Garland Press, 2000.

Clark, Caryl. "Haydn in the Theater: The Operas." In *The Cambridge Companion to Haydn*, ed. Caryl Clark, 176–199. Cambridge: Cambridge University Press, 2005.

Contessa, E. W., Friedrich Baron de la Motte Fouqué, and E. T. A. Hoffmann. *Kinder-Mährchen*. Berlin: G. Reimer, 1839.

Cook, Nicholas. *Analysing Multimedia*. Oxford: Oxford University Press, 1998.

Crangle, Richard, Mervyn Heard, and Ine Van Dooren, eds. *Realms of Light*. London: Magic Lantern Society, 2005.

Crary, Jonathan. *Techniques of the Observer: On Vision and Modernity in the Nineteenth Century*. Cambridge, MA: MIT Press, 1990.

———. "Géricault, the Panorama, and Sites of Reality in the Early Nineteenth Century." *Grey Room* 9 (2002): 5–25.

Cuno, Cosmus Conrad. *Observationes durch dessen verfertigte Microscopia*. Augsburg: Samuel Fincke, 1734.

Czerny, Carl. *Die Kunst des Vortrags der älteren und neueren Klavierkompositionen*. Vienna: A. Diabelli u. Comp., 1846.

———. *On the Proper Performance of All Beethoven's Works for the Piano*, ed. Paul Badura-Skoda. Vienna: Universal, 1970.

Dahlhaus, Carl. *The Idea of Absolute Music*, trans. Roger Lustig. Chicago: University of Chicago Press, 1989.

———. *Nineteenth-Century Music*, trans. J. Bradford Robinson. Berkeley: University of California Press, 1989.

Daston, Lorraine, and Katharine Park. *Wonders and the Order of Nature, 1150–1750*. New York: Zone Books, 1998.

de Celles, François Bedos. *The Organ-Builder*, trans. Charles Ferguson. Raleigh, NC: Sunbury Press, 1977.

DeNora, Tia. *Beethoven and the Construction of Genius: Musical Politics in Vienna, 1792–1803*. Berkeley: University of California Press, 1997.

Dent, Edward J. *The Rise of Romantic Opera*. New York: Cambridge University Press, 1976.

Denton, Eric Hadley. "The Technological Eye: Theater Lighting and *Guckkasten* in Michaelis and Goethe." In *The Enlightened Eye: Goethe and Visual Culture*, ed. Evelyn K. Moore and Patricia Anne Simpson, 239–264. New York: Rodopi, 2007.

de Salins, Pierre Marion. *L'heureuse pêche: comédie pour les ombres à scenes changeantes: représentée pour la premiere fois au Château de ***, le 22 décembre 1767*. Paris: Chez Le Jay, 1770.

Deutsch, Otto. *Schubert: A Documentary Biography*, trans. Eric Blom. New York: Da Capo Press, 1977.

Dobson, Austin. "Ombres Chinoises." *The Cornhill Magazine* 11 (1901): 757–765.

Dolan, Emily. *The Orchestral Revolution: Haydn and the Technologies of Timbre*. Cambridge: Cambridge University Press, 2013.

Dörner, Klaus. *Madmen and the Bourgeoisie*. Oxford: Basil Blackwell, 1981.

Eberhard, Johann Peter. *Abhandlungen vom physikalischen Aberglauben und der Magie*. Halle: Magdeburgischen, 1778.

Eckartshausen, Karl von. *Aufschlüsse zur Magie aus geprüften Erfahrungen über verborgene philosophische Wissenschaften und verdeckte Geheimnisse der Natur*, vol. 2. Munich: Joseph Leutner, 1791.

Elsaesser, Thomas. "Early Film History and Multi-Media: An Archaeology of Possible Futures?" In *New Media, Old Media: A History and Theory Reader*, ed. Wendy Hui Kyong Chung and Thomas Keenan, 13–25. New York: Routledge Press, 2006.

Erlmann, Veit, ed. *Hearing Cultures: Essays on Sound, Listening and Modernity*. New York: Berg, 2004.

Eversberg, Gerd. "'*Ombres chinoises*': Zur Geschichte eines Medienspektakels seit dem siebzehnten Jahrhundert." In *Dei Mobilisierung des Sehens*, ed. Harro Segeberg, 45–68. Munich: Wilhelm Fink Verlag, 1996.

Felski, Rita. "'Context Stinks!'" *New Literary History* 42 (2011): 573–591.

[Flamand-Grétry, L.-V.]. *Théâtre de Séraphin ou des ombres chinoises*, vol. 2. Paris: Desmarest, 1809.

Flax, Neil M. "Goethe's Faust II and the Experimental Theater of His Time." *Comparative Literature* 31 (1979): 154–166.

Florian, [Jean-Pierre Claris de]. *Fables de M. de Florian*. Paris: P. Didot, 1792.

Forster, Edward Morgan. *Howards End*. New York: A. A. Knopf, 1921.

Franklin, Peter. *Seeing Through Music: Gender and Modernism in Classic Hollywood Film Scores*. New York: Oxford University Press, 2011.

Friedberg, Anne. *Window Shopping: Cinema and the Postmodern*. Berkeley: University of California Press, 1994.

———. *The Virtual Window*. Cambridge, MA: MIT University Press, 2006.

Friedlaender, Max. *Das deutsche Lied im 18. Jahrhundert: Quellen und Studien*, vol. 1. Stuttgart and Berlin: J. G. Cotta'sche, 1902.

Funk, Christlieb Benedict. *Natürliche Magie oder Erklärung verschiedner Wahrsager- und Natürlicher Zauberkünste*. Berlin und Stettin: Friedrich Nicolai, 1783.

Füsslin, Georg. *Der Guckkasten: Einblick, Durchblick, Ausblick*. Stuttgart: Füsslin, 1995.

Gehler, Johann Samuel Traugott. *Physikalisches Wörterbuch oder Versuch einer Erklärung der vornehmsten Begriffe und Kunstwörter der Naturlehre*, vol. 3. Leipzig: Schwickert, 1790.

Gerhard, Anselm. *The Urbanization of Opera: Music Theater in Paris in the Nineteenth Century*. Chicago: University of Chicago Press, 1998.

Goehr, Lydia. *The Imaginary Museum of Musical Works: An Essay in the Philosophy of Music*. New York: Oxford University Press, 1992.

Goethe, Johann Wolfgang von. *The Autobiography of Goethe. Truth and Poetry: From My Own Life*, vol. 2, trans. [unnamed]. London: George Bell and Sons, 1882.

———. *Jahrmarktsfest zu Plundersweilern*, ed. Max Herrmann. Berlin: Weidmannsche Buchhandlung, 1900.

———. *Wilhelm Meisters theatralische Sendung*, ed. Hary Mayne. Berlin: J. G. Cotta'sche Buchhandlung Nachfolger, 1911.

———. *Briefwechsel zwischen Goethe und Zelter in den Jahren 1799 bis 1832*. Vol. 20 of *Sämtliche Werke nach Epochen seines Schaffens* (Münchner Ausgabe), ed. Karl Richter et al. Munich: Carl Hanser, 1991.

[Goldoni, Carlo]. *Die Welt im Monde*. Oels: Samuel Gottlieb Ludwig [c. 1750].

Golinski, Jan. "Barometers of Change: Meteorological Instruments as Machines of Enlightenment." In *The Sciences in Enlightened Europe*, ed. William Clark, Jan Golinski, and Simon Schaffer, 69–93. Chicago: University of Chicago Press, 1999.

Gorbman, Claudia. *Unheard Melodies: Narrative Film Music*. Bloomington: Indiana University Press, 1987.

Gotwals, Vernon. *Joseph Haydn: Eighteenth-Century Gentleman and Genius*. Madison: University of Wisconsin Press, 1963.

Gramit, David. *Cultivating Music: The Aspirations, Interests, and Limits of German Musical Culture, 1770–1848*. Berkeley: University of California Press, 2002.

Grey, Thomas. "Tableaux Vivants: Landscape, History Painting, and the Visual Imagination in Mendelssohn's Orchestral Music." *19th-Century Music* 21 (1997): 38–76.

Griesinger, Georg August. *Biographische Notizen über Joseph Haydn*. Leipzig, 1810.

Grimm, Friedrich Melchior. *Correspondance littérarie, philosophique et critique de Grimm et de Diderot, depuis 1753 jusqu'en 1790*, vol. 7. Paris: Furne, 1812.

Grimm, Jacob and Wilhelm. *Deutsches Wörterbuch*, 32 vols. Leipzig: S. Hirzel, 1971.

Grover-Friedlander, Michal. *Vocal Apparitions: The Attraction of Cinema to Opera*. Princeton, NJ: Princeton University Press, 2005.

Gruber, Georg Wilhelm. *Bürger's Gedichte für das Klavier und die Singstimme gesezt*. Nuremberg, 1780.

Guillemain, Charles-Jacob. *Le Pont Cassé*, ed. Denis Bordat and Francis Boucrot. Paris: Editions du Scarabée, 1954.

Guyot, [Edmé-Gilles]. *Neue physikalische und mathematische belustigungen, oder Sammlung von neuen kunststücken, zum vergnügen*, vol. 6, trans. John Christoph Thenn. Augsburg: Eberhard Kletts, 1776.

Guyot, [Edmé-Gilles]. *Nouvelles récréations physiques et mathématiques*, vol. 3, part 6. Paris: Gueffier, 1774.

Halle, Johann Samuel. *Magie, oder die Zauberkräfte der Natur, so auf den Nutzen und die Belustigung angewandt worden*, vol. 1. Wien: Johann Thomas Edlen von Trattnern, 1787.

Hamilton, John T. *Music, Madness, and the Unworking of Language*. New York: Columbia University Press, 2008.

Hankins, Thomas L., and Robert J. Silverman. *Instruments and the Imagination*. Princeton, NJ: Princeton University Press, 1995.

Hayles, Katherine. *How We Became Posthuman*. Chicago: University of Chicago Press, 1999.

Head, Matthew. "Music with 'No Past'? Archaeologies of Joseph Haydn and 'The Creation.'" *19th-Century Music* 23 (2000): 191–217.

Heard, Mervyn. *Phantasmagoria: The Secret Life of the Magic Lantern*. Hastings: Projection Box, 2006.

Heartz, Daniel. *Music in European Capitals: The Galant Style, 1720–1780*. New York: Norton, 2003.

Hecht, Hermann. *Pre-Cinema History: An Encyclopaedia and Annotated Bibliography of the Moving Image before 1896*. London: Bowker-Saur, 1993.

Hegel, G. W. F. *Aesthetics: Lectures on Fine Art*, trans. T. M. Knox. Oxford: Oxford University Press, 1975.

Helmuth, Johann Heinrich. *Volksnaturlehre zur Dämpfung des Aberglaubens*, 4th ed. Braunschweig: Verlag der Schulbuchhandlung, 1798.

Helvétius, Claude Adrien. *De l'esprit; or Essays on the Mind, and its Several Faculties*, trans. William Mudford. London: M. Jones, 1807.

Herbert, Stephen. "Magic Lantern Shows." In *Encyclopedia of Early Cinema*, ed. Richard Abel. New York: Routledge, 2005.

Herder, Johann Gottfried. "*Kalligone* (1800)." In *Musical Aesthetics: A Historical Reader*, vol. 2: *The Nineteenth Century*, ed. Edward A. Lippman, 33–43. New York: Pendragon Press, 1988.

———. "Emanuel Swedenborg: The Greatest Spirit-Seer of the Eighteenth Century," trans. Gregory R. Johnson and Glenn Alexander Magee. *Studia Swedenborgiana* 14 (2005) [online].

———. *Shakespeare*, trans. Gregory Moore. Princeton, NJ: Princeton University Press, 2008.

Hertz, Deborah. "Public Leisure and the Rise of Salons." In *Early Modern Europe: Issues and Inter-*

pretations, ed. James B. Collins and Karen L. Taylor, 282–296. Malden, MA: Blackwell Publishing, 2006.

Hibberd, Sarah. "*Le Naufrage de la Méduse* and Operatic Spectacle in 1830s Paris." *19th-Century Music* 36/3 (2013): 248–263.

Hiller, Johann Adam. "Abhandlung über die Nachahmung der Natur in der Musik." In *Historisch-kritische Beyträge zur Aufnahme der Musik*, vol. 1, ed. F. W. Marpurg, 515–543. Berlin, 1755.

Hirsch, Marjorie. *Schubert's Dramatic Lieder*. Cambridge: Cambridge University Press, 1993.

Hoffmann, E. T. A. "Recension," *Allgemeine musikalische Zeitung* 12 (July 4, 1810).

———. *The Best Tales of Hoffmann*, trans. Alexander Ewing, ed. E. F. Bleiler. New York: Dover Publications, 1967.

———. *E. T. A. Hoffmann's Musical Writings*, ed. David Charlton, trans. Martyn Clarke. Cambridge: Cambridge University Press, 1989.

———. *Fantasy Pieces in Callot's Manner: Pages from the Diary of a Traveling Romantic*, trans. Joseph M. Hayse. Schenectady, NY: Union College Press, 1996.

———. "Master Flea." In *The Golden Pot and Other Tales*, trans. Ritchie Robertson, 239–375. New York: Oxford University Press, 2000.

Holmes, Richard. *The Age of Wonder: How the Romantic Generation Discovered the Beauty and Terror Science*. New York: Pantheon Books, 2008.

Hooke, Robert. *Micrographia: or, Some Physiological Descriptions of Minute Bodies Made by Magnifying Glasses with Observations and Inquiries Thereupon*. London: J. Martyn and J. Allestry, 1665.

Howard, Leslie, perf. *Liszt: The Recitations with Pianoforte* (Hyperion CDA67045, 1996) [CD].

Howes, David. "Hearing Scents, Tasting Sights: Towards a Cross-Cultural Multimodal Theory of Aesthetics." In *Art and the Senses*, ed. Francesca Bacci and David Melcher, 161–181. New York: Oxford University Press, 2013.

Huhtamo, Erkki. "The Pleasures of the Peephole: An Archaeological Exploration of Peep Media." In *Book of Imaginary Media*, ed. Erick Kluitenberg, 74–155. Rotterdam: De Balie, 2006.

———. "Toward a History of Peep Practice." In *A Companion to Early Cinema*, ed. André Gaudreault, Nicolas Dulac, and Santiago Hidalgo, 32–51. New York: John Wiley & Sons, 2012.

———. *Illusions in Motion: Media Archaeology of the Moving Panorama and Related Spectacles*. Cambridge, MA: MIT Press, 2013.

Hunter, Mary. "'To Play as if from the Soul of the Composer': The Idea of the Performer in Early Romantic Aesthetics." *Journal of the American Musicological Society* 58 (2005): 357–398.

Jander, Owen. *Beethoven's "Orpheus" Concerto: The Fourth Piano Concerto in Its Cultural Context*. Hillsdale, NY: Pendragon, 2009.

Johnson, James H. *Listening in Paris: A Cultural History*. Berkeley: University of California Press, 1995.

Jones, Caroline A. *Eyesight Alone: Clement Greenberg's Modernism and the Bureaucratization of the Senses*. Chicago: University of Chicago Press, 2005.

Joubert, Estelle. "Songs to Shape a German Nation: Hiller's Comic Operas and the Public Sphere." *Eighteenth-Century Music* 3 (2006): 213–230.

Joyce, Laurence. "Haydn and Salomon, and Their Visit to Herschel's Telescope at Slough, June 1792." *Haydn Studien* 8 (2000–2004): 289–292.

Julien, Jean-Rémy. *Musique et publicité: Du Cri de Paris . . . aux messages publicitares radiophoniques et télévisés*. Paris: Flammarion, 1989.

Jütte, Robert. *A History of the Senses: From Antiquity to Cyberspace*. Malden, MA: Polity, 2005.

Kaffka, Johann Christoph. *Der Guk Kasten, oder das beste kommt zulezt*. Breslau, 1782.

Kane, Brian. *Sound Unseen: Acousmatic Sound in Theory and Practice*. New York: Oxford University Press, 2014.

Kant, Immanuel. "Dreams of a Spirit-Seer Elucidated by Dreams of Metaphysics (1766)." In *Theoretical Philosophy, 1755–1770*, ed. David Walford, 301–359. Cambridge: Cambridge University Press, 1992.

———. *Anthropology from a Pragmatic Point of View*, ed. and trans. Robert B. Louden. Cambridge: Cambridge University Press, 2006.

———. *Observations on the Feeling of the Beautiful and Sublime and other writings*, ed. and trans. Patrick Frierson and Paul Guyer. Cambridge: Cambridge University Press, 2011.

Kaufman, Thomas. *Painterly Enlightenment: The Art of Franz Anton Maulbertsch, 1724–1796*. Chapel Hill: University of North Carolina Press, 2005.

Keefe, Simon P. "The Concerto from Mozart to Beethoven: Aesthetic and Stylistic Perspectives." In *The Cambridge Companion to the Concerto*, ed. Simon P. Keefe Cambridge: Cambridge University Press, 2005.

Keiser, Reinhard, and C. H. Weidemann. *Le Bon Vivant, oder Die Leipziger Messe*. Hamburg: Gennagel, 1710.

Kenny, Neil. *Curiosity in Early Modern Europe: Word Histories*. Wiesbaden: Harrasscwitz, 1998.

Kerman, Joseph. "Beethoven Quartet Audiences: Actual, Potential, Ideal." In *The Beethoven Quartet Companion*, ed. Robert Winter and Robert Martin. Berkeley: University of California Press, 1994.

———. *Concerto Conversations*. Cambridge, MA: Harvard University Press, 1999.

Kinderman, William. *Beethoven*, 2nd ed. Oxford: Oxford University Press, 2009.

Kirnberger, Johann. *Gesänge am Clavier*. Berlin und Leipzig: George Jacob Decker, 1780.

Kittler, Friedrich. *Discourse Networks 1800/1900*, trans. Michael Metteer. Stanford, CA: Stanford University Press, 1990.

———. *Gramophone, Film, Typewriter*, trans. Geoffrey Winthrop-Young. Stanford, CA: Stanford University Press, 1999.

Klinger, Friedrich Maximillian. *Fausts Leben, Thaten und Höllenfahrt*. St. Petersburg: Johann Friedrich Kriele, 1791.

———. *Faustus: His Life, Death, and Descent into Hell*, trans. George Borrow. London: W. Simpkin and R. Marshall, 1825.

Knittel, K. M. "Pilgrimages to Beethoven: Reminiscences by His Contemporaries." *Music & Letters* 84 (2003): 19–54.

Koepnick, Lutz, and Nora M. Alter, eds. *Sound Matters: Essays on the Acoustics of Modern German Culture*. New York: Berghahn Books, 2004.

Kopp, A. "Schöne Spielewerk, schöne Rarität!" In *Archiv für Kulturgeschichte*, vol. 2, ed. Georg Steinhausen, 296–317. Berlin: Alexander Duncker Verlag, 1904.

Kosmann, J. W. A. "Ueber das Kunstkabinet und die optischen Vorstellungen des Herrn Enslen." In *Denkwürdigkeiten und Tagesgeschichte der Mark Brandenburg*, vol. 2, ed. J. W. A. Kosmann and Th. Heinsius. Berlin: Beliz und Braun, 1797.

Kotzebue, August von. *Neue Schauspiele*, vol. 11. Leipzig: Paul Gotthelf Kummer, 1804.

Kramer, Lawrence. *Music and Poetry, the Nineteenth Century and After*. Berkeley: University of California Press, 1984.

———. "Music and Representation: The Instance of Haydn's *Creation*." In *Music and Text: Critical Inquiries*, ed. Steven Paul Scher, 139–162. Cambridge: Cambridge University Press, 1992.

———. *Classical Music and Postmodern Knowledge*. Berkeley: University of California Press, 1995.

Landon, H. C. Robbins. *Haydn: Chronicle and Works*, 5 vols. Bloomington: Indiana University Press, 1976–1980.

Langan, Celeste. "Understanding Media in 1805: Audiovisual Hallucination in 'The Lay of the Last Minstrel.'" *Studies in Romanticism* 40 (2001): 49–70.

La Nuit de Varennes. DVD, dir. Ettore Scola. São Paulo, Brazil: Versatil Home Video, 2006[1982].

Lastra, James. *Sound Technology and the American Cinema: Perception, Representation, Modernity*. New York: Columbia University Press, 2000.

Le Bar, Ann. "The Domestication of Vocal Music in Enlightenment Hamburg." *Journal of Musicological Research* 19 (2000): 97–134.

Le Guin, Elisabeth. *Boccherini's Body: An Essay in Carnal Musicology*. Berkeley: University of California Press, 2006.

Le Huray, Peter, and James Day, eds. *Music and Aesthetics in the Eighteenth and Early-Nineteenth Centuries*, New York: Cambridge University Press, 1981.

Leppert, Richard. *The Sight of Sound: Music, Representation, and the History of the Body*. Berkeley: University of California Press, 1993.

Lessing, Gotthold Ephraim. *Laocoon*, trans. Robert Phillimore. London: Macmillan, 1874.

———. *Werke und Briefe in zwölf Bänden*, ed. Gunter E. Grimm. Frankfurt am Main: Deutscher Klassiker Verlag, 1997.

Lewald, August. "Musik." In *Europa, chronic der gebildeten welt*, vol. 1. Leipzig and Stuttgart: J. Scheible's Verlags-Expedition, 1835.

———. *Album der Boudoirs*. Leipzig, 1836.

Lippman, Edward A. *A History of Western Musical Aesthetics*. Lincoln: University of Nebraska Press, 1992.

———. "The Tonal Ideal of Romanticism." In *The Philosophy & Aesthetics of Music*. Lincoln: University of Nebraska Press, 1999.

Little, Meredith Ellis. "Forlana." *Grove Music Online. Oxford Music Online*. Oxford University Press, accessed July 18, 2014, http://www.oxfordmusiconline.com/subscriber/article/grove/music/09980.

Little, William Alfred. *Gottfried August Bürger*. New York: Twayne Publishers, 1974.

Loesser, Arthur. *Men, Women, and Pianos: A Social History*. New York: Simon and Schuster, 1954.

Loughridge, Deirdre. "Haydn's *Creation* as an Optical Entertainment." *Journal of Musicology* 27 (2010): 9–54.

———. "Technologies of the Invisible: Optical Instruments and Musical Romanticism." PhD diss., University of Pennsylvania, 2011.

———. "Magnified Vision, Mediated Listening, and the 'Point of Audition' of Early Romanticism." *Eighteenth-Century Music* 10 (2013): 179–211.

———. "Celestial Mechanisms: Adam Walker's Eidouranion, Celestina, and the Advancement of Knowledge." In *Sound Knowledge: Music and Science in London, 1789–1851*, ed. J. Q. Davies and Ellen Lockhart. Chicago: University of Chicago Press, forthcoming.

Lyons, James, and John Plunk, eds. *Multimedia Histories: From the Magic Lantern to the Internet*. Exeter: University of Exeter Press, 2007.

MacIntyre, Bruce C. *Haydn, The Creation*. New York: Schirmer Books, 1998.

Malet, Antoni. "Early Conceptualizations of the Telescope as an Optical Instrument." *Early Science and Medicine* 10 (2005): 237–262.

Maniates, Maria Rika, and Richard Freedman. "Street Cries." *Grove Music Online. Oxford Music Online*. Oxford University Press, accessed July 29, 2014, http://www.oxfordmusiconline.com/subscriber/article/grove/music/26931.

Manicke, Dietrich, ed. *Balladen von Gottfried August Bürger: In Musik gesetzt von André, Kunzen, Zumsteeg, Tomaschek und Reichardt*. Mainz: Verlag B. Schott's Söhne.

Mannoni, Laurent. *The Great Art of Light and Shadow: Archaeology of the Cinema*, trans. Richard Crangle. Exeter: University of Exeter Press, 2001.

Marmontel, Jean-François. *Mémoires*, vol. 3. Paris: Xhrouet, 1804.

———. *Memoirs of Marmontel*, trans. Hugh Murray. Edinburgh: Mundell, Doig, and Stevenson, 1808.

Marmontel, Jean-François, and André-Ernest-Modeste Grétry. *Zémire et Azor: comédie-ballet, en vers, et en quatre actes*. Paris: Vente, 1771.

Martin, Benjamin. *The Young Gentleman and Lady's Philosophy: In a Continued Survey of the Works of Nature and Art by Way of Dialogue*, 3 vols. London: W. Owen, 1772–1782.

Mathew, Nicholas. "Beethoven's Political Music, the Handelian Sublime, and the Aesthetics of Prostration." *19th-Century Music* 33 (2009): 110–150.

———. *Political Beethoven*. Cambridge: Cambridge University Press, 2013.

McKeon, Michael. "Mediation as Primal Word: The Arts, the Sciences, and the Origins of the Aesthetic." In *This Is Enlightenment*, ed. Clifford Siskin and William Warner, 384–412. Chicago: University of Chicago Press, 2010.

McLuhan, Marshall. *Gutenberg Galaxy: The Making of Typographic Man*. Toronto: University of Toronto Press, 1962.

———. *Understanding Media: The Extensions of Man.* New York: McGraw Hill, 1964.

Michaelis, Johann Benjamin. *Operetten*, vol. 1. Leipzig: Dyckischen Buchhandlung, 1772.

Mitchell, W. J. T. "Showing Seeing: A Critique of Visual Culture." *Journal of Visual Culture* 1 (2002): 165–181.

Monelle, Raymond. *The Musical Topic: Hunt, Military, and Pastoral.* Bloomington: Indiana University Press, 2006.

Moyneux, William. *Dioptrica Nova, a Treatise of Dioptricks*, 2nd ed. London, 1709.

Mozart, Leopold. *Versuch einer gründlichen Violinschule.* Wien: C. Stephenson, 1922.

———. *A Treatise on the Fundamental Principles of Violin Playing*, trans. Editha Knocker. New York: Oxford University Press, 1951.

Mozart, Wolfgang Amadeus. *The Letters of Mozart and His Family*, vol. 2, trans. and ed. Emily Anderson. New York: St. Martin's Press, 1966.

Mulvey, Laura. "Uncertainty: Natural Magic and the Art of Deception." In *Death 24x a Second: Stillness and the Moving Image*, 33–53. London: Reaktion Books, 2006.

Mungen, Anno. *"BilderMusik": Panorama, Tableux vivants und Lichtbilder als multimediale Darstellungsformen in Theater- und Musikaufführungen vom 19. bis zum frühen 20. Jahrhundert*, 2 vols. Remscheid: Gardez! Verlag, 2006.

"Music Discovery Still Dominated by Radio, Says Nielsen Music 360 Report," http://www.nielsen.com/us/en/press-room/2012/music-discovery-still-dominated-by-radio—says-nielsen-music-360.html. August 14, 2012.

Nedbal, Martin. "Live Marionettes and Divas on the Strings: *Die Zauberflöte*'s Interactions with Puppet Theater." *Opera Quarterly* 28 (2012): 20–36.

Neefe, Christian Gottlob. *Amors Guckkasten: Eine komische Operette*, ed. Gerhart von Westerman. München: Drei Masken Verlag, 1922.

Newcomb, Anthony. "New Lights on Weber's Wolf Glen." In *Opera and the Enlightenment*, ed. Thomas Bauman and Marita McClymonds, 61–88. Cambridge: Cambridge University Press, 1995.

Nicolai, Christoph Friedrich. "A Memoir on the Appearance of Spectres or Phantoms occasioned by Disease, with Psychological Remarks. Read by Nicolai to the Royal Society of Berlin, on the 28th of February, 1799." *A Journal of Natural Philosophy, Chemistry, and the Arts* 6 (November 1803): 161–179.

November, Nancy. "Instrumental Arias or Sonic Tableaux: 'Voice' in Haydn's String Quartets Opp. 9 and 17." *Music & Letters* 89 (2008): 346–372.

Objects of Vertu and Miniatures: Christie's South Kensington, Tuesday, 7 July 1998 at 10:30 A.M. South Kensington: Christie's, 1998.

Oergel, Maike. *The Return of King Arthur and the Nibelungen: National Myth in Nineteenth-Century English and German Literature.* Berlin: Walter de Gruyter, 1998.

Oetterman, Stephan. *The Panorama: History of a Mass Medium*, trans. Deborah Lucas Schneider. New York: Zone Books, 1997.

Ord-Hume, W. J. G. *Automatic Organs: A Guide to Orchestrions, Barrel Organs, Fairground, Dancehall & Street Organs including Organettes.* Atglen, PA: Schiffer Publishing Ltd., 2007.

Outram, Dorinda. *Panorama of the Enlightenment.* Los Angeles: J. Paul Getty Museum, 2006.

Packer, Randall, and Ken Jordan, eds. *Multimedia: From Wagner to Virtual Reality.* New York: W. W. Norton, 2002.

Parsons, James. "The Eighteenth-Century Lied." In *Cambridge Companion to the Lied*, ed. James Parsons, 35–62. Cambridge: Cambridge University Press, 2004.

Pinetti, Giuseppe. *Physical Amusements and Diverting Experiments.* London, 1784.

Piper, Andrew. "Transitional Figures: Image, Translation, and the Ballad from Broadside to Photograph." In *Book Illustration in the Long Eighteenth Century: Reconfiguring the Visual Periphery of the Text*, ed. Christina Ionescu, 157–194. Newcastle: Cambridge Scholars Publishing, 2011.

Pisko, Franz Josef. *Licht und Farbe: Eine gemeinfassliche Darstellung der Optik*, vol. 2. Munich: R. A. Oldenbourg, 1869.

Piva, Franco, cond. *Galuppi: Il mondo della luna* (Bongiovanni B00000I597, 1997) [CD].

Plantinga, Leon. *Beethoven's Concertos: History, Style, Performance.* New York: Norton, 1999.
Polzonetti, Pierpaolo. *Italian Opera in the Age of the American Revolution.* Cambridge: Cambridge University Press, 2011.
Purschke, Hans. "Puppenspiel und verwandte Künste in der Reichsstadt Nürnberg." *Mitteilungen des Vereins für Geschichte der Stadt Nürnberg* 68 (1981): 221–258.
Quantz, Johann Joachim. *On Playing the Flute*, trans. Edward Reilly. New York: Schirmer Books, 1975.
———. *Versuch einer Anweisung, die Flöte traversière zu spielen: Reprint der Ausgabe Berlin 1752.* München: Deutscher Taschenbuch Verlag, 1992.
Radlof, Johann Gottlieb. *Mustersaal aller teutschen Mund-arten, enthaltend Gedichte, prosaische Aufsätze und kleine Lustspiele in den verschiedenen Mund-arten aufgesetzt*, vol. 2. Bonn: Büschler, 1822.
Reck, von der. "Nachricht von der Philidorschen Geisterbeschwörung." *Berliner Monatsschrift* 13/5 (May 1789): 456–473.
"Récréations Musicales et Pittoresques." *Le Ménestrel* 158 (December 11, 1836): [1–2].
Reichardt, Friedrich. "An junge Künstler." *Musikalisches Kunstmagazin* 1 (1782): 1–7.
Reichardt, Rolf, and Deborah Louise Cohen. "Light against Darkness: The Visual Representation of a Central Enlightenment Concept." *Representations* 61 (1998): 95–148.
Richards, Annette. "Automatic Genius: Mozart and the Mechanical Sublime." *Music & Letters* 80 (1999): 366–389.
———. *The Free Fantasia and the Musical Picturesque.* Cambridge: Cambridge University Press, 2001.
Richardson, John, Claudia Gorbman, and Carol Vernallis, eds. *Oxford Handbook of New Audiovisual Aesthetics.* Oxford: Oxford University Press, 2013.
Richter, Joseph. *Das schöne Milchmädchen, oder Der Guckkasten.* Vienna: Mit von Kurtzbekischen Schriften, 1796.
Riley, Matthew. *Musical Listening in the German Enlightenment: Attention, Wonder, and Astonishment.* Burlington, VT: Ashgate, 2004.
Riley, Terrance. "Composing for the Machine." *European Romantic Review* 20 (2009): 367–379.
Rintel, Wilhelm, ed. *Carl Friedrich Zelter: Eine Lebensbeschreibung.* Berlin: Otto Janke, 1861.
Robinson, David. "Augsburg Peepshows." *Print Quarterly* 5 (1988): 188–191.
Rochlitz, Friedrich. "Der Besuch im Irrenhause." *Allgemeine musikalische Zeitung* 6/39 (June 27, 1804): col. 645–654.
———. "Recension: Lenore, Ballade von G. A. Bürger." *Allgemeine musikalische Zeitung*, 15/41 (October 13, 1813): col. 674–676.
Rogers, Holly. *Sounding the Gallery: Video and the Rise of Art-Music.* New York: Oxford University Press, 2013.
Ronyak, Jennifer. "Performing the Lied, Performing the Self: Singing Subjectivity in German, 1790–1832." PhD diss., Eastman School of Music, 2010.
———. "'Serious Play,' Performance, and the Lied: The Stägemann *Schöne Müllerin* Revisited." *19th-Century Music* 34 (2010): 141–167.
———. "Introduction," Colloquy: "Studying the Lied: Hermeneutic Traditions and the Challenge of Performance." *Journal of the American Musicological Society* 67 (2014): 543–549.
Ross, Alex. "Schubert on the Beach: The New World Symphony's Radical New Home," *The New Yorker* (February 14, 2011) [online].
Rossell, Deac. "Enslen, Johann Carl (1759–1848)." In *Encyclopedia of Nineteenth-Century Photography*, vol. 1, ed. John Hannavy, 491–493. New York: CRC Press, 2007.
———. *Laterna magica/Magic Lantern*, vol. 1. Stuttgart: Füsslin Verlag, 2008.
Rousseau, Jean-Jacques. "*Dictionnaire de musique.*" In *Music and Aesthetics in the Eighteenth and Early-Nineteenth Centuries*, ed. Peter le Huray and James Day, 85–94. New York: Cambridge University Press, 1988.
Rowland, David. *A History of Pianoforte Pedalling.* Cambridge: Cambridge University Press, 1993.

Rumph, Stephen C. *Beethoven after Napoleon: Political Romanticism in the Late Works*. Berkeley: University of California Press, 2004.

Russell, John. *A Tour in Germany, and Some of the Southern Provinces of the Austrian Empire, in the Years 1820, 1821, 1822*, vol. 2. Edinburgh: Archibald Constable, 1825.

Schaffer, Simon. "Natural Philosophy and Public Spectacle in the Eighteenth Century." *History of Science* 21 (1983): 1–43.

Schellenberg, Renata. "The Impact of Ossian: Johann Gottfried Herder's Literary Legacy." In *The Voice of the People: Writing the European Folk Revival, 1760–1914*, ed. Matthew Campbell and Michael Perraudin, 9–20. New York: Anthem Press, 2012.

Schenker, Heinrich. *Der Tonwille: Pamphlets/Quarterly Publications in Witness of the Immutable Laws of Music*, vol. 2, ed. William Drabkin. New York: Oxford University Press, 2004[1921–1924].

Schlegel, August Wilhelm von. *Kritische Schriften*, vol. 2. Berlin: G. Reimer, 1828.

Schleuning, Peter. "Die Fantasiermaschine: Ein Beitrag zur Geschichte der Stilwende um 1750." *Archiv für Musikwissenschaft* 27 (1970): 192–213.

Schmitt, Ulrich. *Revolution im Konzertsaal: zur Beethoven-Rezeption im 19. Jahrhundert*. New York: Schott, 1990.

Schön, Eric. *Verlust der Sinnlichkeit, oder, Die Verwandlungen des Lesers: Mentalitätswandel um 1800*. Stuttgart: Klett-Cotta, 1987.

Schubart, Christian Friedrich. "Auserlesene scherzhafte und zärtliche Lieder, in Musik gesetzt von Johann André," *Deutsche Chronik auf das Jahr 1775*, Stück 3 (January 9, 1775): 22–23.

———. "Lenore, von Bürger, in Musik gesetzt von André." *Deutsche Chronik auf das Jahr 1775*, Stück 63 (August 7, 1775): 502–503.

Schulz, Johann Abraham Peter. *Lieder im Volkston*, vol. 2. Berlin: Georg Jacob Decker, 1785.

Scola, Ettore, dir. *Il mondo nuovo/La Nuit de Varennes*. Triumph Releasing Corporation, 1982.

Schwarzkopf, J. v. "Erzählung einer neulichen Geisterzitation in Berlin." *Berliner Monatsschrift* 13/5 (May 1789): 474–484.

Sebastian, Thomas. "Schattenspiel und laterna magica als poetologische Metaphern—Antizipation der Traumfabrik bei Justinus Kerner und Achim von Arnim." *Michigan Germanic Studies* 23 (1997): 1–23.

Senner, Wayne M., Robin Wallace, and William Meredith, eds. *The Critical Reception of Beethoven's Compositions by His German Contemporaries*, 2 vols. Lincoln: University of Nebraska Press, 1999.

Séraphin, François Dominique. *Feu Séraphin: Histoire de ce spectacle depuis son origine jusqu'a sa disparition, 1776–1870*. Lyon: N. Scheuring, 1875.

Shapin, Steven, and Simon Schaffer. *Leviathan and the Air-Pump: Hobbes, Boyle, and the Experimental Life*. Princeton, NJ: Princeton University Press, 2011.

Shaw-Miller, Simon. *Visible Deeds of Music: Art and Music from Wagner to Cage*. New Haven, CT: Yale University Press, 2002.

Singy, Patrick. "Huber's Eyes: The Art of Scientific Observation before the Emergence of Positivism." *Representations* 96 (2006): 54–75.

Sisario, Ben. "What's Billboard's No. 1? Now YouTube Has a Say." *New York Times*, February 21, 2013, C3.

Sisman, Elaine. "Haydn's Solar Poetics: The *Tageszeiten* Symphonies and Enlightenment Knowledge." *Journal of the American Musicological Society* 66 (2013): 5–102.

———, ed. *Haydn and His World*. Princeton, NJ: Princeton University Press, 1997.

Smart, Mary Ann. *Mimomania: Music and Gesture in Nineteenth-Century Opera*. Berkeley: University of California Press, 2004.

Smollett, Tobias George. *The Adventures of Ferdinand Count Fathom*, 2 vols. London, 1753.

Spalding, Keith, ed. *An Historical Dictionary of German Figurative Usage*, 6 vols. Oxford: B. Blackwell, 1952–.

Speaight, George. *The History of the English Puppet Theater*. Carbondale: Southern Illinois University Press, 1990.

Stafford, Barbara Maria. *Body Criticism: Imaging the Unseen in Enlightenment Art and Medicine*. Cambridge, MA: MIT Press, 1993.

———. *Artful Science*. Cambridge, MA: MIT Press, 1996.

Stafford, Barbara Maria, and Frances Terpak, eds. *Devices of Wonder: From the World in a Box to Images on a Screen*. Los Angeles: Getty Research Institute, 2001.

Steinberg, Michael P. *Listening to Reason: Culture, Subjectivity, and Nineteenth-Century Music*. Princeton, NJ: Princeton University Press, 2006.

Sterne, Jonathan. *The Audible Past: Cultural Origins of Sound Reproduction*. Durham, NC: Duke University Press, 2003.

Stowell, Robin. *Beethoven: Violin Concerto*. Cambridge: Cambridge University Press, 1998.

Striggio, Alessandro. *Orfeo: Favola in Musica*, trans. Gilbert Blin (Boston Early Music Festival, 2012), http://www.bemf.org/media/1213_opera_libretto.pdf.

Sulzer, Johann Georg. "Tone Painting." In *Aesthetics and the Art of Musical Composition in the German Enlightenment: Selected Writings of Johann Georg Sulzer and Heinrich Christoph Koch*, ed. Nancy Kovaleff Baker and Thomas Christensen. Cambridge: Cambridge University Press, 1995.

Taruskin, Richard. *The Oxford History of Western Music*, vol. 2: *Music in the Seventeenth and Eighteenth Centuries*. Oxford: Oxford University Press, 2005.

Telemann, Georg Philipp. *24 Oden*. Hamburg: Christian Herold, 1741.

Temperley, Nicholas. *Haydn: The Creation*. New York: Cambridge University Press, 1991.

Thayer, Alexander Wheelock. *Thayer's Life of Beethoven*, rev. and ed. Elliot Forbes. Princeton, NJ: Princeton University Press, 1967.

"Theatralische Gaukelspiele in Berlin: Berlin den 20ten Juni 1796." *Journal des Luxus und der Moden* 11 (1796): 422–429.

Thomas, Downing A. *Aesthetics of Opera in the Ancien Régime, 1647–1785*. Cambridge: Cambridge University Press, 2002.

Thompson, Emily. *The Soundscape of Modernity: Architectural Acoustics and the Culture of Listening in America, 1900–1933*. Cambridge, MA: MIT Press, 2002.

Tieck, Ludwig. *William Lovell: Ein Briefroman aus den Jahren 1793–1796*. Darmstadt: Schriftenreihe Agora, 1961.

Tolley, Thomas. *Painting the Cannon's Roar: Music, the Visual Arts, and the Rise of an Attentive Public in the Age of Haydn, c. 1750 to c. 1810*. Burlington, VT: Ashgate, 2001.

Tovey, Donald. *Essays in Musical Analysis*. London: Oxford University Press, 1981.

Traber, Zacharias. *Nervus opticus sive tractatus theoricus*. Vienna, 1675.

Tresch, John. "Estrangement of Vision: Edgar Allen Poe's Optics." In *Observing Nature—Representing Experience: The Osmotic Dynamics of Romanticism, 1800–1850*, ed. Erna Fiorenti. Berlin: Reimer, 2007.

———. "The Prophet and the Pendulum: Sensational Science and Audiovisual Phantasmagoria around 1848." *Grey Room* 43 (2011): 16–41.

———. *The Romantic Machine: Utopian Science and Technology after Napoleon*. Chicago: University of Chicago Press, 2012.

Triest, Johann Karl Friedrich. "Bermerkungen über der Ausbildung der Tonkunst in Deutschland im achtzehnten Jahrhundert." *Allgemeine musikalische Zeitung* 3 (March 11, 1801): col. 409, trans. Susan Gillespie as "Remarks on the Development of the Art of Music in Germany in the Eighteenth Century," in *Haydn and His World*, ed. Elaine Sisman. Princeton, NJ: Princeton University Press, 1997.

Van Helden, Albert, and Thomas L. Hankins. "Introduction: Instruments in the History of Science." *Osiris* 9 (1994): 1–6.

Vernallis, Carol. *Unruly Media: YouTube, Music Video, and the New Digital Cinema*. Oxford: Oxford University Press, 2013.

Vernallis, Carol, Amy Herzog, and John Richardson, eds. *The Oxford Handbook of Sound and Image in Digital Media*. Oxford: Oxford University Press, 2013.

Voskuhl, Adelheid. *Androids in the Enlightenment*. Chicago: University of Chicago Press, 2013.

Wackenroder, Wilhelm Heinrich. *Wilhelm Heinrich Wackenroder's Confessions and Fantasies*, trans. Mary Hurst Schubert. University Park: Pennsylvania State University Press, 1971.

Wackenroder, Wilhelm Heinrich, and Ludwig Tieck. *Phantasien über die Kunst*. Hamburg, 1799.

Watkins, Holly. *Metaphors of Depth in German Musical Thought: From E. T. A. Hoffmann to Arnold Schoenberg*. New York: Cambridge University Press, 2011.

Weber, Carl Maria von. "Ueber die Oper, Undine." *Allgemeine musikalische Zeitung* 19/12 (March 19, 1817): col. 204.

———. *Carl Maria von Weber: Writings on Music*, ed. John Warrack, trans. Martin Cooper. Cambridge: Cambridge University Press, 1981.

Webster, James. "The Rhetoric of Improvisation in Haydn's Keyboard Music." In *Haydn and the Performance of Rhetoric*, ed. Tom Beghin and Sander M. Goldberg. Chicago: University of Chicago Press, 2007.

Wegner, Georg Wilhelm. *Schauplatz Vieler Ungereimten Meynungen und Erzehlungen*, vol. 1. Berlin and Leipzig: Ambrosius Haude, 1735.

Weis, Friedrich Wilhelm. *Lieder mit Melodien II*. Lübeck, 1776.

Weisse, Christian Felix. *Komische Opern*, vol. 3. Carlsruhe: C. G. Schmieder, 1778.

———. *Christian Felix Weissens Selbstbiographie, herausgegeben von dessen Sohne Christian Ernst Weisse und dessen Schwiegersohne Samuel Gottlob Frische, mit Zusätzen von dem Letztern*. Leipzig: Georg Voss, 1806.

Wellbery, David E. *Lessing's Laocoon: Semiotics and Aesthetics in the Age of Reason*. Cambridge: Cambridge University Press, 1984.

———. *The Specular Moment: Goethe's Early Lyric and the Beginnings of Romanticism*. Palo Alto, CA: Stanford University Press, 1996.

Werrett, Simon. *Fireworks: Pyrotechnic Arts and Sciences in European History*. Chicago: University of Chicago Press, 2010.

Weston, Helen. "In the Eye of the Artist." In *Realms of Light*, ed. Richard Crangle, Mervyn Heard, and Ine Van Dooren. London: Magic Lantern Society, 2005.

Wiegleb, Johann Christian. *Die natürliche Magie: aus allerhand belustigenden und nützlichen Kunststücken bestehend*, vol. 3. Berlin and Stettin: Friedrich Nicolai, 1779.

Wiener, Philip P. "Leibniz's Project of a Public Exhibition of Scientific Inventions." *Journal of the History of Ideas* 1 (1940): 232–240.

Will, Richard. *The Characteristic Symphony in the Age of Haydn and Beethoven*. Cambridge: Cambridge University Press, 2002.

Wilson, Catherine. "Visual Surface and Visual Symbol: The Microscope and the Occult in Early Modern Europe." *Journal of the History of Ideas* 49/1 (1988): 85–108.

Wolf, Georg Friedrich. *Lieder mit Melodien*. Nordhausen, 1781.

Wood, Gillen D'Arcy. *The Shock of the Real: Romanticism and Visual Culture, 1760–1860*. New York: Palgrave, 2001.

Woodmansee, Martha. *The Author, Art, and the Market: Rereading the History of Aesthetics*. New York: Columbia University Press, 1994.

Wraxall, N. W. *Memoirs of the Courts of Berlin, Dresden, Warsaw, and Vienna in the Years 1777, 1778, and 1779*, vol. 1. London: Cadell, 1800.

Zammito, John H. *The Genesis of Kant's Critique of Judgment*. Chicago: University of Chicago Press, 1992.

Zelter, Carl Friedrich. "Die Schöpfung: Ein Oratorium." *Allgemeine musikalische Zeitung* 4 (March 10, 1802): col. 389.

———. *Darstellungen seines Lebens*, ed. Johann-Wolfgang Schottländer. New York: Georg Olms Verlag, 1978.

"*Zémire & Azor, comedie-ballet*." *Journal helvétique; ou Annales littéraries et politiques de l'Europe* (January 1772): 33–42.

Ziolkowski, Theodore. *German Romanticism and Its Institutions*. Princeton, NJ: Princeton University Press, 1992.

Zohn, Steven. "Images of Telemann: Narratives of Reception in the Composer's Anecdote, 1750–1830." *Journal of Musicology* 21 (2004): 459–486.

Zucchi, John E. *Little Slaves of the Harp: Italian Child Street Musicians in Nineteenth-Century Paris, London, and New York*. Buffalo: McGill-Queen's University Press, 1992.

INDEX

Page numbers in bold indicate images or score excerpts.